WITHDRAWN

Hurricane Katrina and the Redefinition of Landscape

Hurricane Katrina and the Redefinition of Landscape

DeMond Shondell Miller and
Jason David Rivera

LEXINGTON BOOKS

A division of
ROWMAN & LITTLEFIELD PUBLISHERS, INC.
Lanham • Boulder • New York • Toronto • Plymouth, UK

LEXINGTON BOOKS

A division of Rowman & Littlefield Publishers, Inc.
A wholly owned subsidiary of The Rowman & Littlefield Publishing Group, Inc.
4501 Forbes Boulevard, Suite 200
Lanham, MD 20706

Estover Road
Plymouth PL6 7PY
United Kingdom

Copyright © 2008 by Lexington Books

All rights reserved. No part of this publication may be reproduced, stored in a retrieval system, or transmitted in any form or by any means, electronic, mechanical, photocopying, recording, or otherwise, without the prior permission of the publisher.

British Library Cataloguing in Publication Information Available

Library of Congress Cataloging-in-Publication Data

Miller, DeMond Shondell, 1973–
  Hurricane Katrina and the redefinition of landscape / DeMond Shondell Miller and Jason David Rivera.
    p. cm.
  Includes bibliographical references and index.
  ISBN-13: 978-0-7391-2146-7 (cloth : alk. paper)
  ISBN-10: 0-7391-2146-4 (cloth : alk. paper)
  1. Hurricane Katrina, 2005. 2. Landscape changes—Louisiana—New Orleans. 3. Landscape changes—Louisiana—Gulf Coast. 4. New Orleans (La.)—Environmental conditions. 5. Gulf Coast (La.)—Environmental conditions. I. Rivera, Jason David, 1983– II. Title.
  HV636 2005 .N4 M55 2008
  307.3'416—dc22   2007048353

Printed in the United States of America

∞™ The paper used in this publication meets the minimum requirements of American National Standard for Information Sciences—Permanence of Paper for Printed Library Materials, ANSI/NISO Z39.48–1992.

We wish to dedicate this book to the people of Louisiana, Mississippi, Alabama, and Florida.

Contents

Figures	ix
Tables	xi
Photospreads	xiii
Acknowledgments	xv
Introduction	1
1 The Concepts of Place and Landscape	8
2 The Physical Landscape	24
3 The Cultural and Economic Landscapes	46
4 The Political Landscape	67
5 Views of Changing Landscapes	91
6 Civic Trust	106
Conclusion	126
Bibliography	146
Index	169
About the Authors	171

Figures

Figure 4.1   Ineffective Leadership
Figure 4.2   The Budget Motel
Figure 7.1   The Traditional View
Figure 7.2   The Modern View
Figure 7.3   The Future View

Tables

Table 3.1   New Orleans' Racial Composition
Table 3.2   Average Household Incomes
Table 3.3   New Orleans' Poverty Rate

Photospreads

First

Figure 1.1	The Disaster Landscape
Figure 1.2	Views of a Disaster Landscape
Map 2.1	A Map of Louisiana and of the River Mississippi
Map 2.2	Plan of New Orleans the Capital of Louisiana; With the Disposition of its Quarters and Canals as They Have Been Traced by Mr. de la Tour in the Year 1720
Map 2.3	Hardee's New Geographical, Historical, and Statistical Map of Louisiana
Map 2.4	Plan of the City and Suburbs of New Orleans from an Actual Survey Made in 1815
Map 2.5	Plan of the City and Environs of New Orleans, Taken From an Actual Survey
Map 2.6	Topographical and Drainage Map of New Orleans and Surroundings

Second

Utter Destruction
Obliteration
A view of the Lower 9th Ward
Hazards in the disaster landscape
Now Hiring with Bonus
Disturbed Cemetery Sites
The disturbed dead
Steps Leading Nowhere
Indiscriminant Destruction
Where a House Once Stood

Acknowledgments

From DeMond Shondell Miller

The opportunity to create this type of book would not be possible without the encouragement of my parents, who spent countless hours traveling with me as I attended events throughout my primary years to today as they joined in my quest to understand the social realities that define daily life in southern Louisiana. Secondly, I wish to thank my all of my former instructors and the professional staff who were instrumental in the development of this manuscript. Finally, I would like to acknowledge my Grandmother, Mrs. Myrtis Ruth Bell Craft, whose strength, resolve, and faith in God has sustained her throughout the recovery and rebuilding process and inspires our entire family.

From Jason David Rivera

I would like to thank my family and friends for their continuous support during the writing of this book. I would also like to thank Cindy for her effort in editing and formatting this book, which was an immense help. Special thanks to all of the individuals and organizations that contributed pictures and graphics to this book, making it a more complete piece. Lastly, I would like to thank Joel Yelin and Emiliano Serrano for their research support in the composition of this piece.

Human activities are transforming the global environment, and these global changes have many faces: ozone depletion, tropical deforestation, acid depletion, and increased atmospheric gasses that trap heat and may warm the global climate. For many of these troubling transformations, the data and analyses are fragmentary, scientific understanding is incomplete, and the long-term implications are unknown. Yet even against a background of uncertainty, it is abundantly clear that human activities now match or even surpass natural processes as agents of change in the planetary environment (Silver & DeFries, 1990, p. iii).

Introduction

Natural disasters of significant size that result in destruction and disorder will always be beyond the capacity of conventional measures for understanding and coping. Natural disasters often impinge upon the "stability"—actual or perceived—of the social, political, and historical order. Disaster produces questioning and anxiety; in its wake are people poised to abandon the values of the past or, conversely, to cling to the old with renewed vigor.

—Ogasawara (1999, p. 172-173)

It is rare in modern history to find the complete destruction of all, or nearly all, of an industrialized city in the aftermath of a natural disaster. Even after the San Francisco earthquake in April 1906, large sections of the city remained open for business and governance while other parts of the city were inoperable (Comfort, 2006). Likewise, after the Great Chicago Fire of 1871, whole neighborhoods remained functional, allowing communication, governance, and other aspects of civil life to carry on with their daily life patterns. However, with the flooding of approximately 80 percent of New Orleans in 2005, the entire city shut down. During the immediate aftermath of Hurricane Katrina, New Orleans, Louisiana, a major metropolitan city in the United States, "went offline" leaving the entire city inoperable and rendering services, such as communications, water, electrical power, sewer, gas distribution, port distribution, and transportation infrastructures, nonfunctional (Comfort, 2006). A series of compounded actions from the past, including levee construction, gradual coastal erosion, and more recent actions taken days prior to the hurricane's arrival, jeopardized life, liberty, and property, and the effects of the actions on the people and their connection to the physical landscape was, and still is, profound.

This work is presented in such a way that we gather information vital to understanding how individual events come together to set the stage for the human tragedy. These events resulted in secondary and tertiary disasters in the shadow of America's first mega-disaster of the 21st century—Hurricane Katrina. By acting as a bridge to the literature on community and environmental sociology, our work explores how place attachment to a landscape symbolizes a connection to the natural environment. This book acknowledges that landscapes are symbolic environments created by humans that confer meaning on nature and give the environment definition and form through a particular angle of vision and special beliefs and values (Greider & Garkovich, 1994). Landscapes, by their very nature, reflect our self-definitions that are grounded in socio-cultural phenomena that can transform a grassy knoll into a suburban enclave with all of the amenities a society values, which symbolically represents the collective culture of its inhabitants. According to Burley, Jenkins, and Azcona (2006), "the concept of landscape encompasses the social elements of place attachments. Thus landscapes reveal how physical and natural features become imbedded with social and communal meaning" (p. 22).

This book is concerned with the role of people's attachment to a place, and ultimately a place within a physical landscape, that results in their connection to society at large. We explore how different landscapes contribute to overall redevelopment in the aftermath of disaster, specifically natural disasters. The issue at hand is understanding the lived topography[1] as viewed through the daily lives of those who remained behind or returned after Hurricane Katrina. To this end, the constitution of the physical, cultural, economic, and political landscapes serve as the basis for understanding the context in which people connect physically, socially, and emotionally to their ecological surroundings. To support this discussion, we consider the juxtaposition of the landscapes and how survivors navigate through them as they cope with abrupt changes. This adaptation is fueled by one's symbolic connection with the local ecology.

When we attempt to understand the consequences of radical changes taking place in the natural environment after a disaster, we must be mindful of the multiple landscapes and the socio-cultural contexts in which they develop. Therefore, when the environment is suddenly in disarray caused by natural calamity, the disruption of the lived topographical experience has profound consequences. Before a disaster, during times of normalcy, attachment to place is not in the forefront of consciousness. Post-disaster attachment to a place advances to the front of conscious thought due to loss or perception lost (Brown & Perkins, 1992).

City in Ruin

In late August 2005, America and the world were transfixed by the images of a submerged cityscape. Scenes of a city 12 feet under water, flat slabs of foundation where houses once stood, boats floating down streets, weary soldiers, and chaos in Faubourg Marigny and other parts of the city all created an indelible image of the city that many will never forget. Shortly after the storm, the Lower 9th Ward section of the city, in the shadow of the ruptured Industrial Canal, became the focal point of the city's state of ruin, symbolic of so much of what had gone wrong. When the waters finally receded in the Lower 9th Ward, what remained was a community ravaged by flood waters, filled with vacant lots where complete houses once stood.

By the time thousands of people were rescued from their once familiar terrain, the landscape became synonymous with sorrow. It was a place where they had lost so much that the memories attached to the New Orleans landscape only echoed their plight to survive amid a surreal landscape of floating corpses. The city became a place where the roofs looked more like islands than the tops of houses as residents broke through their roofs and displayed signs calling for help in their time of abandonment. Just as the streets, public parks, private residences, and downtown lay in disarray, New Orleans' Mayor, Ray C. Nagin, and Louisiana Governor, Kathleen Blanco, blamed the social catastrophe unfolding in the streets and the institutional, whole-scale failure of the system on the Federal Emergency Management Agency's (FEMA's) inability to quickly mobilize and mount an appropriate response.

The Disaster Landscape

The term *disaster landscape* is used to differentiate between an area hit by a natural disaster, such as Hurricane Katrina, and the normal physical landscape. The disaster landscape is used to clearly illustrate the difference between a pre-impact and a post-impact place. Embodied in this concept is the notion of sudden change and newness. Sudden change takes place in such a way that the survivors initially have difficulty comprehending how the disaster took place and then have difficulty dealing with the total devastation left behind so that the landscape can be returned to its pre-disaster form. The term disaster landscape also implies a "new normalcy" in that the rules that once structured and governed life no longer match the current state of existence. In the case of New Orleans, the current state of existence approximately 1 year after the disaster challenges the notion of the pre-disaster landscape and what is normal.

Although the changes in the physical environment are perceived, those that remain reflect the social aspects of the disaster landscape as well. For instance, anthropologist Paul Bohannan (1995) classified disasters into two categories: physical disasters, such as earthquakes and floods, and social disasters, such as famines and wars. He argued that physical disasters are unlike social disasters because they do not appreciably alter cultural systems. Social disasters disrupt behavior while simultaneously introducing a *new* culture, or a disaster subculture. Thus it follows, according to Burley et al. (2006), that the "identification with place may serve as the basis for action in accordance with place" (p. 39), and action is structured by the disaster.

Disaster as a Structuring Event

For the purpose of this discussion, an extension is proposed of the concept of disaster used by Fritz (1961) and Barton (1970). Fritz argued that a disaster is any event "concentrated in time and space in which a society or a relatively self-sufficient subdivision of society undergoes severe danger and incurs losses to its members and physical appurtenances [so] that the social structure is disrupted and the fulfillment of all or some of the essential functions of the society is prevented" (p. 665). Barton's definition is similar, but it focuses on the social system, contending that "Disasters exist when members of a social system fail to receive expected conditions of life from the system" (p. 38).

Disaster researchers are representative of a wide variety of specialized fields, adding to the difficulty of defining the "core concept" of a disaster (Aday & Ito, 1989). When most scientists discuss disaster, they refer to the actual event in terms of the physical impacts of or problems associated with unplanned, socially disruptive events (Barton, 1970; Dynes, 1970; Fritz, 1961). However, it is not just the physical aspects that are important in defining a disaster. Disasters involve an "abrupt transition from the mundane, relatively safe life into an environment of chaos and hell" (Silverstein, 1992, p. 3). The victim is thrust into a world of falling rubble, fire, toxic gasses, earthquakes, hurricanes, or even gunfire. Whatever the cause of the turmoil, the person is merely fighting to survive.

The key to understanding any disaster is not the event but rather whether the event alters an existing relationship between an individual and groups, disrupts community norms, and compels us to alter our notion of where humans fit in the cycle of life. The distortion of the physical landscape presents a new world and a new reality to victims in the community. When a community is exposed to toxins, the way people view themselves and the world around them changes. Edelstein's (2003) argument of the lifescape challenges the notion that all aspects of life return to normal after a disaster. The symbolic interpretative process is just as important as the changes in cognition that occur after a disaster.

We define *disaster* as a naturally occurring or human-induced event that causes severe damage to the surrounding environment in which agents of society incur physical, social, economic, and/or psychological damages, resulting in a disruption of the routine interactions, ultimately leading to a failure in the existing social network. The failure in the existing or "normal" social order is soon replaced during the initial traumatization of the disaster by a new set of norms that govern daily existence.

The loss of a home, community, and security during a natural disaster such as Katrina only magnifies the loss of a sense of place. After the disaster, one saw homes shattered and smelled decaying animal carcasses. The area was no longer recognizable; in many cases, photographs taken prior to the hurricane were used to assist relief workers and visitors in an attempt to reconnect with the place. Entire communities were destroyed to the point of unrecognizability, and a new or changed landscape—a disaster landscape—replaced the prior landscape. Because such large-scale disasters are rare, it is difficult to integrate the social changes transpiring in each community that result from them. The changes caused by such disasters challenge existing notions of what is to be expected (i.e., hurricanes in paradise), exceeding one's ability to comprehend the changes and impeding recovery. Victims of a disaster have socially constructed realities that conflict with the physical reality of a new set of environmental conditions, yet disasters have a significant place and value in nature. Rolston (1988) is somewhat hesitant about incorporating these natural disasters in his appreciation of the ecosystem because some are so massive and rare that ecosystems have no adaptations to them. Hence, he considers these phenomena anomalies that challenge the general paradigm of nature's landscapes and have an essential beauty (Saito, 1998).

When we embarked on the development of this manuscript, Hurricane Katrina was 6 months behind us and beginning to be a memory of the past for many. As other national priorities became more pressing, the plight of those still rebuilding their lives moved from the center of media focus, loosing their command on the attention of the American public. Following the initial attention given to Katrina by the national media, the spotlight was only focused on New Orleans at the post-disaster update 6 months after Katrina and, thereafter, only when there was a crime or a political event, such as the Mayor's election. Many commentators reported on the recovery process with stories featuring the National Guard, the Army Corps of Engineers, and the citizens' rebuilding efforts; however, few media stories throughout the recovery period covered the impact the landscape had on the city, state, and region's road to recovery. The stories that did appear in the media were those that spoke of abrupt landscape changes in areas where there was mostly infrastructure damage and disruption.

This book seeks to go beyond infrastructure damage to view the physical landscape and how people are attached to it vis-à-vis an ecologic-symbolic approach. However, to only consider one's attachment to the physical landscape, rooted in attachment to place, would oversimplify our current understanding of how place attachment impacts redevelopment in the aftermath of a disaster. The issues involving redevelopment are so complex that we use the landscape approach to help explain the constellation of issues that surround redevelopment.

This work affords us the opportunity to address how the lived topography influences other landscapes layered atop it. In fact, Greider and Garkovich (1994) maintain that "any physical place has the potential to embody multiple landscapes, each of which is grounded in the cultural definitions of those who encounter that place. Every river is more than just one river. Every rock is more than just one rock" (p. 2). Although we do not suggest that the layers we propose in our work are the totality of the disaster landscape, they do offer insight into the role that interaction with the physical landscape has in shaping the development and redevelopment of the other landscapes. We do not negate the importance of the psychological landscape or the emotional landscape, but rather the focus here will be on those landscapes that are more social, economic, and political in nature and that govern behavior.

In this study, we explore the cultural, economic, and political landscapes. The cultural and economic landscapes communicate how the culture and economy of New Orleans are interlinked. More specifically, we view the role of a natural disaster in redistributing wealth and economic opportunity. We argue that the culture and economy are fused to the point where *culture is business* in New Orleans. We contend that the political landscape is vital to the understanding of the layered landscape; again, we do not argue that politics, culture, and economics are the only three variables that determine life in the wake of a disaster, nor do we ignore the psychological underpinnings of behavior within a political landscape. We also argue that the need to be attached to a place and the development of a sense of place is rooted in cultural and economic redevelopment. Furthermore, we discuss past interactions that certain groups have experienced with the environment and society, such as social, economic, and class inequity, and how those experiences affect the formation of positive or negative conceptions of a place.

At the core of this book is the concept of loss. For many, the losses included one's home, way of life, familiar surroundings, and landscape, which is so profound that some people have failed to fully recover from the environmental assault that day. The road to recovery is further complicated by a landscape that remains in disrepair, which further prolongs the survivor's ability to adapt to a new identity as a survivor. When individuals place so much emphasis in their attachment to their surroundings—and they are forced to temporarily leave their surroundings—feelings of severe loss develop. Bridget Dugan, a nursing student and former New Orleans resident, discussed her experiences of surviving Katrina and her current feelings toward her lost city:

> I feel I am slowly forgetting the city I love and was home to me . . . homes are in ruins, neighbors are eerily quiet, and there is no life. There are no birds. The homes are scarred with the paint from the writing of rescue workers showing whether bodies were recovered or not. As hard as you try, the paint does not completely come off. Paint does not disguise the painful memory. (Dugan, 2007, p. 45)

Dugan (2007) also stated that it is important to recognize that a temporary or permanent loss of culture can create an identity crisis for an individual. A loss of culture is exactly what New Orleans may experience if survivors are not a part of the rebuilding process. Because New Orleans has such a rich cultural heritage, a loss of its diversity would be empirically disheartening and would permanently damage the social, economic, and cultural landscape of the city that once existed. For the survivors, the difficult task of processing the initial event's trauma and rescue period, while simultaneously reconnecting to the newly repaired landscape in the aftermath of Hurricane Katrina, is aided when strategic steps are taken to improve the physical surroundings, decrease vulnerability for future disasters, and create public policies, all of which ensure a steady rate of recovery. As for the returning survivors, they must forge ahead to reconnect with their landscape, make new memories, and strengthen the bonds with a city that poses an ever-present risk to their property, sense of place, security, and lives. Dugan concluded by stating that "your life can change in a day or even in a moment. It takes time to come out of the shock and begin to process the horror of what happened to you so suddenly and in such a short span of time. I believe crisis can promote growth and new levels of consciousness" (p. 45-46).

The discussion of the dynamics of these landscapes, with specific geographical reference to New Orleans and the Gulf Coast, unfolds in this book. Through discussion of the past dynamics of these landscapes, we will explain their relationships and their interdependence on one another. These landscapes, which have been divided for discussion purposes, have significant influences over the development of one another. After understanding the dynamics of these landscapes, we will discuss how Hurricane Katrina altered these dynamics and ultimately changed the culture known by previous generations.

It is our intent to describe the changes unfolding and present ways social theorists come to understand those changes. After explaining the social changes that have taken place as well as those that will most likely take place as a result of the altered landscapes, we discuss the interactional past and the interactional potential of much of the region. We will then discuss the concept of trust in government and how the local and national governments can foster better relations among the population in the areas affected by Hurricane Katrina. Here, we propose models for citizen participation in policy construction and its implications for civic trust, sustainable development, and social equality.

Finally, we will discuss the lessons that have been learned from Hurricane Katrina, including those about the government's inadequacies. Our work seeks to make the public aware of the recurring social dynamics behind the implementation of disaster policy and how the hurricane has devastatingly affected the future of New Orleans and the Gulf Coast.

Note

1. The term *topography* refers to the practice, art, and science of picturing place, either through written description or pictorial representation. The goal of topography is to accurately portray place, which means it prepares for the construction of an accurate mental image—the picture of a place in one's mind. In this instance, the term *lived* refers to the experiential aspect of one's relationship with his or her surroundings to create a lifeworld. In essence, the lifeworld is the world as it is immediately experienced through one's everyday life (Backhaus, 2005).

Chapter 1
The Concepts of Place and Landscape

There was a Japanese elm in the courtyard . . . it used to blossom in the springtime. They were destroying that tree, the wrecking crew. We saw it together. She asked the man whether it could be saved. No, he had [a] job to do and was doing it. I screamed and cried out. The old janitor, Joe, was standing out there crying to himself. . . . At night, the sparrows used to roost in those trees and it was something to hear the singing of those sparrows. All that was soft and beautiful was destroyed. You saw no meaning in anything anymore. There is a college campus there now.

—Florence Scala (Studs, 1968, p. 32)

The spaces people inhabit profoundly impact the molding of their individuality. Whether one person or a community of people live in a specific place, the aesthetics of the landscape and the symbols used by groups to represent them influence the behaviors of the community and those who reside nearby. According to Read (1996), "the way in which [humans] actually and symbolically create landscape within the cultural community probably reflects other organizing principles of that society and its world view" (p. 2). In many ways, the physical landscape affects our way of life and our interpretations of the world. As Cosgrove (1989) pointed out, "All landscapes are symbolic...reproducing cultural norms and establishing the dominant groups across all of a society" (p. 125). By understanding the changes in the physical landscape of an area after a natural disaster, people are capable of appreciating the workings of nature (Miller, 2006a; Saito, 1998) and its ability to change the landscape. Through the transcendence of time, people are reminded that they, and their societies and communities, are still subject to the forces of nature.

After a natural disaster passes, the abrupt and total physical destruction alters the physical landscape and changes the society and its individuals. Specifically, Fried (1963) characterized this disruption as the consequent perception of that place as it inevitably changes. This study follows Fried's definition of disruption:

> Any severe loss may produce a disruption in one's relationship to the past, to the present, and the future. Losses generally bring about fragmentation of routines, of relationships, and of expectations, and generally imply an alteration of the world of physically available objects and spatially oriented action. It is a disruption on the sense of continuity which is an ordinarily taken-for-granted framework for functioning in a universe which has temporal, social, and spatial

dimensions ... *the loss of an important place represents a change in a potentially significant component of the experience of continuity* [emphasis added]. (p. 153)

It is the alteration of a population's perception of the world around them that influences their perceptions about themselves, their community, and society. Hamerton (1885, as cited in Riley, 1992) contended that people's moods and personalities correspond with specific, real-world, mood-inducing landscapes. Places have the ability to influence our understanding of the world and indirectly guide us in the establishment of specific moral views about the world around us. For Tuan (1991), nature first had to be conceptualized as a place before it could be appreciated:

> that the "quality" of place is more than just aesthetic or affectional, that it also has a *moral* dimension, which is to be expected if language is a component in the construction and maintenance of reality, for language—ordinary language—is never morally neutral. (as cited in Bird, 2002, p. 544)

Therefore, when a landscape is abruptly changed by a natural disaster such as Hurricane Katrina, there is an abrupt change in the way inhabitants who have lived in an area their entire lives relate to the landscape. Human relationships with a landscape change as their experiences and memories attained in that place are altered, which may have positive or negative results (Riley, 1992). In short, when a place is indistinguishable from another because it is littered with debris, the "place" has the potential to disappear. According to Huigen and Meijering (2005), "Places disappear and lose their identity when they are no longer distinguished from their context, when they vanish in their context" (p. 29). This may happen when places become part of a larger whole. Hence, when topographic features are indistinguishable, humans have a difficult time connecting to place.

Changes in the Landscape

Whether altered by natural causes (erosion) or manmade transformations (demolition and construction), the landscapes of an area inevitably change over time. Built urban landscapes are similar to natural landscapes; Kyvig and Marty (2000, p. 165) present a six-step cycle through which the urban landscape changes:

(1) construction
(2) abandonment
(3) conversion
(4) abandonment
(5) demolition
(6) new construction

Sometimes this cycle skips one step, but its essential motion persists; construction, abandonment, and eventual new construction are found wherever humans are present. Built landscapes' processes such as abandonment and destruction are directly influenced by environmental factors and urban planning, which take time to complete. Progression through the cycle and natural environmental changes allows the population to adjust to its new surroundings gradually, but when the destruction is sudden, like that caused by Hurricane Katrina, the progressive psychological adjustment needed by people does not have time to occur. In the advent of a massively destructive natural disaster, the physical landscape changes literally overnight, which, in turn, abruptly changes or displaces the cycle (Kyvig & Marty, 2000; Miller & Rivera, 2006a). According to Brown and Perkins (1992) the slow process of becoming attached to a place can be disrupted quickly and create a long-term phase of dealing with loss in addition to repairing or recreating attachments to places, which is why disruptions to the continuity of a landscape's evolution are potentially devastating to the place attachment indicative to it.

Because natural disasters challenge current notions of what is to be expected in the normalcy of everyday activities, these events exceed our ability to readily comprehend their majesty and overall place in nature. Humans have a tendency to disregard disasters as "freak occurrences" that do not usually have long-lasting effects on society as a whole; however, it is this disregard for such events that makes the disasters all the more dangerous in their destruction of places and their subsequent influences on society.

When a disaster causes extreme devastation to the environment and makes significant alterations to the physical landscape of an area, perceptions about its long-term effects tend to change. In addition to the perceptional and physical alterations, major disasters disrupt normative patterns (Brown & Perkins, 1992; Fried, 1963). When an extreme change in the physical landscape occurs, significant changes in perception about the place occur that are inherently different from those of the past generations who lived there. According to Day (2002) and Miller and Rivera (2006a), communities understand places, and that understanding is built on memories of the past.

Changes to a place threaten the interpretation of past memories. The physical landscape of a place allows the resident population to project their values and culture into the future so following generations may adhere to similar values (Foote, 1988; Foote, 1997; Lowenthal, 1985). If a landscape is significantly changed, the place attachment of past generations is impossible to duplicate in future generations because specific features of the landscape are now missing (Milligan, 1998). Just as people who first come in contact with one another make first impressions, one is also made when a person first comes in contact with a place. This first impression is pivotal to understanding and interpreting place (Day, 2002). Significantly altered places inspire a variety of new perceptions about the place relative to past generations and communities native to the site.

When natural disasters indiscriminately destroy communities, some physical structures that were present prior to the devastation are not reconstructed. Failure to rebuild certain objects once prominent in society symbolizes a community's belief that some of the aspects of society are not "wanted" or are "wished to be forgotten." Therefore, the elements reconstructed symbolize the cultural values that the society seeks to highlight (Foote, 1997). The use and reuse of a specific place aids in the construction and instruction of the contemporary community residing within it (Basso, 1996). However, through the transmission of values in reconstruction, the decision to rebuild specific elements of the old community is a societal and political statement on the behalf of the forces in charge of reconstruction (Baker, 2003). Because communities rely on places to transfer knowledge and symbolically represent values from one generation to the next, it is important to conceptualize how people and communities "bond" with their surrounding landscapes.

The Importance of Place

Cochrane (1987, as cited in Brown & Perkins, 1992) believed that "Place . . . means permanence, security, nourishment, a center or organizing principle" (p. 281). Although place is understood as a physical location, there is a reciprocal relationship between people and a place, "an interlocking system in which the people and place define one another" (p. 281). Eyles (1989) explained place as follows:

> a centre of felt value, incarnating the experience and aspirations of people. Thus it is not only an arena for everyday life . . . [it also] provides meaning to that life. To be attached to a place is seen as a fundamental human need and, particularly as home, as the foundation of our selves and our identities. Places are thus conceived as profound centres of human experience. As such, they can provide not only a sense of well-being but also one of entrapment and drudgery. To be tied to one place may well enmesh a person in the familiar and routine. (as cited in and Brown and Perkins, 1992, p. 281)

Further, Basso (1996) argued that places are as much a "part of us as we are a part of them" (p. xiv). Places are significant in the development of culture and shared bodies of knowledge within a specific culture:

> sense of place may assert itself in pressing and powerful ways, and its often subtle components—as subtle, perhaps as absent smells in the air or not enough visible sky—come surging into awareness.
>
> It is then we come to see that attachments to places may be nothing less than profound, and that when these attachments are threatened, we may feel threatened as well. Places, we realize, are as much a part of us as we are of them, and senses of place—yours, mine, and everyone else's—partake complexly of both.
>
> And so, unavoidably, senses of place also partake of cultures, of shared bodies of "local knowledge" with which persons and whole communities render their places meaningful and endow them with social importance. (Basso, p. xiii-xiv)

Through shared knowledge, we as communities find places meaningful and attribute special significance to them. For example, native Alaskans living within the Arctic Circle use local knowledge and local building materials to construct homes or igloos. This knowledge, which is particular to its place, has a system that has been passed down throughout the community's history as a means to provide families with shelter from the cold. "The language of landscape rediscovers the dynamic connection between place and those who dwell there"

(Spirn, 1998, p. 17). People are taught how to sustain themselves by understanding a place and its specific characteristics (Chapman, 1979).

Another aspect of place is its ability to evoke meaning and build a socially constructed reality within a woven fabric of dialogues that are enduring and ephemeral. According to Berger and Luckman (1967), the reality of everyday life is an ordered reality. Reality exists in our environment in prearranged patterns before individuals experience them. Thus, places we occupy are intersubjective worlds shared by others in the environment. This type of retrospective world-building is called *placemaking* (Basso, 1996). Placemaking involves the historical construction of "what happened here" and why it is relevant. It includes the social aspects that develop in a given geographical location—how humans arrange themselves at a given site and why the arrangement is successful or unsuccessful. The landscape fabric forges this narrative composed of landscape elements and features that have stories shaped over hundreds of years. Every place has ongoing stories that are recognizable (Spirn, 1998).

For sociologists, the basic organizing concept is a social system similar to the concept of ecosystem (Harper, 2003). Social systems reside in places that lack cultural dimensions. Within the socio-cultural system, there exists a social network of independent individuals, organizations, subsystems, and institutions that remain stable over time. They illustrate cultural patterns distinguishable from other social systems at other locations. One primary difference in subsistence cultures is that they are centered around the fact that human systems are dependent on the environmental changes made by seasonal changes. In socio-cultural systems, everything is ultimately related in a complex social dynamic; one disruption in any part of the system poses potentially devastating consequences for the other parts.

Place and "Space as a Production"

Henri Lefebvre, a French Marxist, existentialist philosopher, and sociologist, began writing about the production of space more than two decades ago. Lefebvre's work provides a framework that can be used to relate the sense of place encountered in cultural landscapes to the political economy (Hayden, 1996). In fact, Lefebvre's (1991) work argues that every society in history has shaped a distinctive social space that meets its intertwined requirements for economic production and social reproduction. "Lefebvre insists that space is *lived* before it is *perceived* and *produced* before it can be *read*, which raises the question of what the virtue of readability actually is, especially because the 'spaces made (produced) to be read are the most deceptive and tricked-up imaginable'"[1] (Dear, 1997, p. 54). A great deal of his work deals with space in and around the body, space of housing, or the space in cities for public use, but in all cases, Lefebvre's work suggests that the production of space is essential to the inner workings of the political economy; his work sees the commonalities between tract houses, identical suites in corporate skyscrapers, and identical shops in malls. In essence, the "cookie cutter" places in contemporary society foster a sense of "placelessness" (Hayden, 1996).

Lefebvre's approach to space and how a place is created provides a framework for constructing important social histories about a place. In his analysis, this approach correlated the social use of the physical landscape and how other types of production, such as economic, political, and cultural, are tied to social reproduction as it is rooted in the physical landscape. One example Lefebvre uses to illustrate the importance of space in shaping social reproduction is the tactic of majority groups restricting the political rights and economic growth of minority groups by limiting their access to space. In many cases, Hayden (1996) maintains that Lefebvre uses spaces as political territories or bounded spaces with some form of enforcement of boundaries. Whereas "Social life structures territory . . . and territory shapes social life" (Dean & Walch, 1989, p. 4).

For instance, "Ghettos and barrios, internments camps and Indian reservations, plantations under slavery and migrant worker camps should all be looked at as political territories and the customs and laws governing them are seen as enforcement of territories" (Hayden, 1996, p. 23). Moreover, political divisions split the landscape in such a way that it is difficult for individuals from the same city but different neighborhoods to describe the city in which they live. For example, Lynch's (1960) studies of mental images of a cityscape suggests that when residents from different communities within a city are asked to draw maps of their community or give directions they have completely different cognitive maps of the city in which they reside. Although there are some limitations of Lynch's work, Jameson's aesthetic of a cognitive mapping approach helps us understand how individuals develop a heightened sense of place for their local environment and provides examples of how this cognitive mapping can raise political consciousness (Jameson, 1991).

Places

According to Basso (1996), prior to the advent of literacy, places served mankind as durable symbols of past events and invaluable aids for remembering and imagining the past. In *Wisdom Sits in Places*, Basso discusses the importance of "place making," which he defined as the ordinary way individuals begin to think about particular places and develop a way of understanding what happened there. He wrote, "it is a common response to common curiosities—what happened here? Who was involved? What was it like? Why should it matter?" (Basso, p. 5).

Whether through the use of monumental architecture, murals and other forms of public art, or the natural physical landscape of a specific area, places aid in the community's transference of values and culture to future generations and outsiders. Specifically, places create an atmosphere of a shared "collective conditioned consciousness" that reinforces individual identities, giving the people and the place the same identity (Bow & Buys, 2003; Relph, 1976). Places have an incorporating ability because they act as the backdrop for human activity and interpersonal involvement (Bow & Buys, 2003). Place-specific human activity transforms space into a place. Whereas "*space* is more abstract than *place*. What begins as undifferentiated space becomes place as we get to know it better and endow it with value" (Tuan, 1977, p. 6), which means that a space becomes a place "as we get to know it better" (Milligan, 1998, p. 5). Furthermore, places are "especially meaningful spaces" (Milligan, 1998, p. 5) because of peoples' ability to associate with or endow sentiment to them. According to Read (1996), "People respond individually to locality, then, and the culture with which they are familiar helps to enlarge, diminish, shape, or transform it. Sense[s] of belonging are allied to attachment and love" (p. 3).

Place theory describes the placemaking process as inherent in identifying specific geographic locations; without some degree of meaning bestowed on it by people, a site does not exist, or at least the location is not a place (Milligan, 1998). It is only when an association with the space is made through resources such as sight, discussion, or even literature that the location becomes a place. Without knowing about a site, it is simply a space no one knows about; however, when people know about the place in some manner, it has the potential to become a place (Milligan, 1998). People's ability to "bond" with places occurs through the interplay of emotions, knowledge, beliefs, behaviors, and actions in relation to a particular place (Altman & Low, 1992; Bonnes & Secchiaroli, 2003; Bow & Buys, 2003; Proshansky, Fabian, & Kaminoff, 1983). According to Norberg-Schulz (1980, as cited in Groat, 1995), places emphasize the quality of a person's existential existence or their being in the world. Although places are experienced by everyone who comes in contact with them, a person develops his or her own concept of a place's identity because experiences differ from person to person (Milligan, 1998). Additionally, because people have different experiences, a place can be represented as the intersection or association of three constituent elements: actions, conceptions, and physical environment (Groat, 1995).

Place identity is the emotional attachment an individual has with a place through the symbolic importance the individual puts on it, which gives meaning and purpose to life (Bow & Buys, 2003; Williams & Vaske, 2002). In this framework, place identity is different from place attachment because, according to Milligan (1998), "*Place attachment* refers to an emotional bonding to a site that decreases the perceived substitutability of other sites for the one in question" (p. 6-7). The concept of place identity develops when the bonds[2] between people and a place become components of self-identity (Bow & Buys, 2003; Proshansky et al., 1983), which occurs through an increase in the sense of community indicative to that place (Bow & Buys, 2003; Relph, 1976). For example, in a local urban setting, a grocery store can exist as an important symbol of a neighborhood and act as a symbol of community identity (Godkin, 1980).

Once created, the loss of a symbol causes a stressful period of disruption followed by a post-disruption phase that is characterized by coping with the lost symbol and creating new ones (Brown & Perkins, 1992). According to Godkin (1980), physical images that are present in places usually evoke specific feelings within individuals (i.e., mountains stir feelings of power and fires create feelings of helplessness). The presence of distinctive physical features, such as mountains or fires, aids in the creation of community and the place's identity in the minds of both the inhabitants and outsiders. It is a process by which the community and the place are defined, and they rely on each other for their identities and perceptions. Place and the memory of a specific place allow humanity to form psychological and social attachments with the built and natural environments and the cultural landscape.

Place Attachment in the Disaster Landscape

In psychological and social-psychological literature, place attachment is acknowledged as a significant factor in the development of self-identity (Godkin, 1980; Searles, 1960; Wenkart, 1961). According to Brown and Perkins (1992):

> Place attachment involves positively experienced bonds, sometimes occurring without awareness, that are developed over time from the behavioral, affective, and cognitive ties between individuals and/or groups and their sociophysical environment. These bonds provide a framework for both individual and communal aspects of identity and have both stabilizing and dynamic features. (p. 284)

Although it is commonly agreed that places have a great effect on an individual and a community's identity, there is no agreed-on understanding as to the process by which the images of place are interwoven into an individual's sense of self (Godkin, 1980). However mysterious the process may be, Milligan (1998) explained that place attachment for an individual occurs as follows:

> first, [as] a site becomes known (space → place), then, to the extent that a known site becomes an object to which an individual is emotionally bonded, as opposed to one that is simply known about, the site becomes one to which a place attachment has been formed (place → place attachment). (p. 7)

The measurement of place attachment is observed through the substitutability of other places for the one in question; the higher the number of substitutes the lower the degree of place attachment (Milligan, 1998; Williams, Patterson, & Roggenbuck, 1992).

Furthermore, place attachment develops through two interrelating processes: *interactional past* and *interactional potential*. Interactional past is the process by which a place acquires meaning for a person over time. Fundamentally, it is the collective experiences (memories of those experiences) that have occurred within the physical boundaries of that place. The interactional potential of a place is created by the site's shape, constraints, and influences over the activities that are perceived as able to happen within its boundaries (Milligan, 1998). Disasters alter the interactional potential of a place, which, in turn, creates unpleasant experiences and memories of that place; these new memories conflict with prior memories, resulting in a *new* interactional past completely different in comparison to the *old* interactional past of that same place.

The interactional past is important to placemaking because the experiences people acquire within places contribute to their attachment with a place. Because placemaking and attachment occur unconsciously (Miller & Rivera, 2007a; Schneekloth & Shibley, 1995), individuals' degree of place attachment varies, which is a byproduct of variance in actual experience, further enhancing the variance of attachment from individual to individual. According to Buttimer (1980):

> The nature of a person's social relationships predisposes him to attach different significance to the routes taken, to the nodes at which interaction occurs, and to places associated with particular events and circumstances. (p. 26)

It is the variation in emotional significance that enables place attachment and identity to change over time. Milligan (1998) describes people's ability to fuse time and experience together in the framework of nostalgia, which created a way in which individuals are able to recognize continuity and organize past experiences by categorizing them to site- and time-specific memories. Moreover, place memory can be used to trigger social memory through the urban landscape. According to Casey (1987), place memory "is the stabilizing persistence of place as a container of experiences that contributes so powerfully to its intrinsic memorability. An alert and alive memory connects spontaneously with place, finding in it features that favor and parallel its own activities. We might say that memory is place-oriented or at least place-supported" (p. 186-187). For example, monuments in cemeteries create a sense of place that relies on a cemetery's interactional past (Miller & Rivera, 2006b) and triggers memories of a forgotten era or an event that defined an era, such as World War II. Through the use and reuse of monuments in a specific cemetery, builders of past monuments are able to convey past ideas to the living. Although the experiences people may currently have in the cemeteries where these monuments are located are undoubtedly different, the builders of the past have attempted to add a sample of their experiences, which may have occurred thousands of years ago, to the minds of individuals today.

Monuments in any setting are an attempt by people to recreate an experience in their lives in which they interacted with others in the place the monuments are found. Place attachment among today's population, although different from past generations, is still molded and influenced by the use of monuments because they evoke an emotional response from the observer. Furthermore, the ability of different generations to perceive the same place differently over time contributes to place attachment. Perception of a place is not fixed within society because the interactional past is constructed based on current situations. Therefore, the interactional past of a specific site may be added to or reinterpreted over the course of several generations (Katovich & Couch, 1992; Katovich & Hintz, 1997; Maines, Sugrue, & Katovich, 1983; Mead, 1929; Milligan, 1998; Zerubavel, 1996).

Unlike the interactional past, the interactional potential of a place is directly linked to the specific physical features of a site (Milligan, 1998). Although the interactional past is based on the interactional potential, the physical conditions of the site limit what is likely to occur there (Maines et al., 1983; Milligan, 1998). In this respect, the physical features of place influence the experiences that can occur at a specific site, thereby influencing the development of an individual's place attachment. Therefore, the interactional potential of a site has a direct effect on the social behavior of individuals, and sometimes entire groups (Mead, 1934; Wilson, 1980), because it dictates, to a certain extent, the experiences that can occur at a given site. Limitations on experiences, then, have the ability to affect an individual's "satisfaction" with a place, influencing the degree of place attachment felt by an individual (Bow & Buys, 2003). For example, both the interactional past and interactional potential of New Orleans' black communities, from its colonization through the events that occurred after Hurricane Katrina, had varying influences on these individuals' place attachment. In New Orleans, issues of interactional past and interactional potential are complicated beyond physical limitations due to the implementation of social controls that created an interactional past. The interactional potential of New Orleans' black communities is limited by the physical features of the city, such as the Gulf of Mexico, Lake Pontchartrain, and the surrounding swamps, in addition to social segregation.

Social segregation contributes to interactional potential because it limits the physical size and places in which blacks citizens can reside. Furthermore, the areas to which blacks were segregated in the past were usually situated in undesirable areas where aesthetics were not pleasing. This socially enforced interactional past and interactional potential has had varying effects on the black communities' attachment to the city. The degree in attachment can be observed in some of the city's residents' lack of desire to move back to New Orleans after being relocated in the aftermath of Katrina. Using Milligan's (1998) model, the individuals' degree of place attachment to New Orleans is relatively low because of their ability to substitute their place attachment to their homes in New Orleans for the cities in which they were relocated.

Disasters of Historic Proportions Reconfiguring the Landscape

Through humans' historical direct and indirect experiences with disasters, several relatively recent disasters have become a permanent part of national and international collective memory. Disasters of historic proportions have the potential to force groups to address not only the vulnerabilities associated with natural hazards but also how competing social, economic, and political powers reshape the physical landscape following major disasters. The rebuilding of the landscape reflects those social aspects that help survivors well beyond the immediate impact phase recreate a sense of place. The following three cases are examples of the disasters that have reshaped the landscapes and humanity's ability to reconnect with its environment after a disaster.

The San Francisco Earthquake and Fire

A violent series of shakes and rumblings occurred just before dawn on April 18, 1906, in San Francisco, California. Within a few seconds, these rumblings set in motion the worst natural disaster America had ever recorded up to that point in time and confounded the scientists of the early 20th century. The strength and magnitude of the earthquake ruptured 296 miles (477 kilometers) of the San Andrea fault, from San Juan Bautista in the northwest to the triple junction at Cape Mendocin, causing massive geographical changes by creating large horizontal displacements and disfiguring large spans of land (Ellsworth, 1990, as cited in U.S. Geological Survey, 2006). The earthquake was felt as far north as Oregon, as far south as Los Angeles, and as far east as central Nevada.

Just as with Hurricane Katrina, hundreds of thousands of people were left homeless and displaced after the earthquake. At the time, 375 deaths were reported (Bronson, 1996), but contemporary approximations place the number of casualties at 3,000 (Ellsworth, 1990, as cited in U.S. Geological Survey, 2006), most of which were in the San Francisco and the San Francisco Bay areas. Other parts of the region experienced extensive damage and landscape alteration. In addition to the earthquakes, San Francisco suffered a series of fires simultaneously that destroyed a large portion of the city. "Fires sprang from gas mains, wood stoves, and toppled lanterns, and church bells were set clanging. Gold-Rush era cisterns were useless, and all but one water main was ruptured" (Galvin, 2003, p. 11). With no water and the city in disarray, civic panic ensued, and the landscape was in complete turmoil; aftershocks from the earthquakes served to further traumatize the citizens and alter the landscape. Nearly two dozen aftershocks continued to rock the city, and fires burned for days after the initial earthquake. Contemporary newspapers wrote a special earthquake edition. The *Call=Chronicle=Examiner* announced, "Death and destruction have been the fate of San Francisco. Shaken by a temblor at 5:13 o'clock yesterday morning, the shock lasting 48 seconds, and scourged by flames that raged diametrically in all directions, the city is a mass of smoldering ruins" (Galvin, 2003, p. 12). The event was summed up by Kurutz, the curator for Special Collections at the California State Library, who stated "It was the great cataclysmic event in twentieth-century California history. As they say, on April 18, 1906, the earth shook [and] the sky burned" (as cited in Galvin, 2003, p. 12).

Before the earthquake, San Francisco was one of the ten largest cities in the United States. Moreover, its cultural and economic influences over the area were imperative parts of the city's identity as a booming industrial port city. However, just as with the change in landscape linked with the natural disaster, the fires rendered the landscape a charred wasteland, a place of dreary and smoky black charcoal that stretched over approximately 3,000 acres and destroyed all of the communities and infrastructures needed to support life. Landon, an eyewitness to the fire, asserted that "an enumeration of the buildings destroyed would be a directory of San Francisco . . . not in history has a modern imperial city been so completely destroyed. San Francisco is gone. Nothing remains of it but memories and a fringe of dwelling-houses on its outskirts" (as cited in Galvin, 2003, p. 12). The damage was estimated to cover over 500 downtown city blocks, but this does not begin to account for the economic, social, and cultural losses that occurred due to these damages.

In 1906, San Francisco was inhabited by approximately 390,000 residents. Reports from 1906 calculate the economic loss from the earthquake at nearly $80 million, with the combined losses of both the earthquake and fire estimated at approximately $400 million (Steinbrugge, 1982). In modern terms, the economic loss factored by 100 corresponds approximately to $40 billion in total fire and earthquake loss and $8 billion in earthquake only losses (Poland, 2006).

Almost immediately after the earthquake, plans were prepared to rebuild the city, partly in an attempt to get the city on its feet sooner rather than later but also in an attempt to repair the cityscape for the 1915 international exposition. During this time, the Army built several thousand relief houses to accommodate the numerous displaced individuals. Much like the Federal Emergency Management Agency trailer villages that continue to span the New Orleans' landscape more than 24 months after Hurricane Katrina, the cottages built in the San Francisco area were overcrowded. With a peak population of 16,448, these cottages were all occupied by 1907. The average cost of a cottage ranged from $100 to $741, and families could rent the cottage for $2.00 per month, which went toward the purchase price of $50.00 per structure (Evans, 2006).

As political and civic leaders rushed to restore the city's functioning, the famed urban planner David Burnham advanced the ideas to rebuild the city and build it "better." Burnham's vision of a neoclassical civic center complex, wider streets, a subway under the main artery, and a more people-friendly Fisherman's Wharf were realized in the remaking of the landscape after the earthquake. New neighborhoods emerged from the ashes while other groups sought land to build more impressive communities such as the neighborhood of Pacific Heights.

The Southeast Asian Tsunami

The Indian Ocean tsunami, caused by the major 9.0 magnitude Sumatra-Andaman megathrust earthquake on December 26, 2004, destroyed entire communities and villages and all the land vegetation; it flattened houses, washed out farmland, and obliterated roads and the people's way of life. For the most part, the attention was focused on the large scale coastal flooding that occurred in six countries: Sri Lanka, India, Indonesia, Bangladesh, Malaysia, and Thailand. In Thailand, among the most effected provinces were Ranong, Phang-rga, Thuket, Krubi, Trang, and Satun (Ratanasermpong & Polngam, 2006). It is estimated that approximately 128,000 people were killed.

> In Thailand alone, 5,393 people were killed, 8,457 injured and 3,062 missing, approximately 58,000 people or 12,000 households affected, 4,800 houses destroyed wholly or partially, 5,000 fishing villages affected, 6,000 fish vessels destroyed. The environment was also greatly affected; marine and coastal parks were damaged; some coral reefs were also destroyed. In addition, coastal flood plain, which is mostly narrow, caused damage to buildings, road networks, bridges, bays or inlets, coastline, etc. Moreover, electricity supply and telephone lines were disrupted for a couple of days. (Ratanasermpong and Polngam, 2006, p. 1)

Major changes to the landscape occurred along the coastline, especially in Sumatra, the Thai Peninsula (Ko Phuket and Khao Lak), Sri Lanka, and Southeastern India. The damage to the land has been estimated to cover an area almost five times the size of the Brussels-capital region of Belgium (European Commission Joint Research Center, n.d.). In total, over 10,960 hectares of forest, 4,393 hectares of the beach, 56,249 hectares of mixed agriculture and village, and 9,193 hectares of urban land were damaged.[3] The 4,393 hectares of barren land lost were mainly sandy beaches although some bare soil areas a few hundred meters inland were badly eroded, severely compromising the ability of people to connect with the land. The altered landscape has presented the risks associated with human development so near to the ocean's shore.

The Galveston Hurricane of 1900

The deadliest hurricane[4] to strike the United States in the 20th century made landfall on the city coast of Galveston, Texas, on September 8, 1900, with prevailing winds estimated at 135 miles per hour. Similar to New Orleans, Galveston is a city that was built near water that surrounds most of the city.[5] Galveston rested on a giant sand bar along the Texas coast that extended into the Gulf of Mexico. The economy and social life were dependent on the water that surrounded it. During the 19th century, the city was known as a prosperous harbor city and was the largest city in Texas.

Before the disaster occurred, Isaac Cline, Director of the Galveston Weather Station,[6] wrote a two-page article in the *Galveston News* (July 15, 1891) where he stated "It would be impossible for any cyclone to create a storm wave which could materially injure the city" (as cited in Heidorn, 2000). In this expert's opinion, the largely prosperous environment of the city and the continual state of development gave city officials and residents a false sense of security. This security was ultimately destroyed, in the same manner as the city, nine years later when the city became a disaster landscape as a result of a direct hit of a hurricane and the subsequent flooding. Because Cline's report was so specific about the nature of a storm and the impossibility of such devastation, city officials ignored any attempts to consider the natural risks associated with previous Gulf storms and hurricanes and did not take sufficient measures to build storm walls that would stave off hurricane storm surges. Moreover, according to Heidorn (2000), naturally occurring sand dunes were cut and used as landfills as a part of the island's residential and commercial development. As human intervention and human landscape alteration increased, so did the level of vulnerability to environmental hazards. The development patterns, false sense of security, and a general sentiment of nonemergency preparedness led to a large amount of landscape change and human suffering. In Cline's (1900) often cited report, he made some detailed observations about the storm and its destruction to personal property in addition to its alterations of the physical landscape. Cline summarized the impact of the storm in the following way:

> Sunday, September 9, 1900, revealed one of the most horrible sights that ever a civilized people looked upon. About three thousand homes, nearly half the residence portion of Galveston, had been completely swept out of existence, and probably more than six thousand persons had passed from life to death during that dreadful night. The correct number of those who perished will probably never be known, for many entire families are missing. Where 20,000 people lived on the 8th, not a house remained on the 9th, and who occupied the houses may, in many instances, never be known. On account of the pleasant Gulf breezes, many strangers were residing temporarily near the beach, and the number of these that were lost can not yet be estimated. I enclose a chart, fig. 2, which shows, by shading, the area of total destruction. Two charts of this area have been drawn independently: one by Mr. A.G. Youens, inspector for the lo-

cal board of underwriters, and the other by myself and Mr. J.L. Cline. The two charts agree in nearly all particulars, and it is believed that the chart enclosed represents the true conditions as nearly as it is possible to show them. That portion of the city west of Forty-Fifth Street was sparsely settled, but there were several splendid residences in the southern part of it. Many truck farmers and dairy men resided on the west end of the island, and it is estimated that half of these were lost, as but very few residences remain standing down the island. For two blocks, inside the shaded area, the damage amounts to at least fifty percent of the property. There is not a house in Galveston that escaped injury, and there are houses totally wrecked in all parts of the city. All goods and supplies not over eight feet above floor were badly injured, and much was totally lost. The damage to buildings, personal, and other property in Galveston County is estimated at above thirty million dollars. The insurance inspector for Galveston states that there were 2,636 residences located prior to the hurricane in the area of total destruction, and he estimates 1,000 houses totally destroyed. The value of these buildings alone is estimated at $5,500,000 (Cline, 1900, p. 374, as cited in National Oceanic & Atmospheric Administration, 2004).

Larson (1999) describes similar accounts of the storm:

All over Galveston, freakish things occurred. Slate fractured skulls and removed limbs. Venomous snakes spiraled upward into trees occupied by people [escaping the flood waters]. [And] a rocket of timber killed a horse in midgallop. (p. 202)

The account by Cline is similar to other communities suffering mass devastation. Yet again, we contend that although the particular environmental conditions differ, extreme environmental change due to disaster alters the connections survivors have with the landscape and that rebuilding and recovery[7] are key elements that effect the re-establishment of one's connection to place.

Summary

Though there has been a great deal of discussion about how people and communities interpret their surroundings, and the ways those perceptions influence individual and community identities, this chapter has only limitedly discussed the social landscapes that develop at a specific site. Because the interactional past and interactional potential influence the identity and cultural dynamics of the resident population, they also mold the creation of the social landscapes in human culture, which include the natural and manmade aesthetic landscape, the cultural, economic, and political landscapes. These landscapes are significantly influenced by the specific place in which they develop because they are the social culmination of that specific culture.

Hurricane Katrina's effect on the physical landscape of New Orleans and the Gulf Coast has significantly influenced the changes in and creation of future social landscapes in the region. Prior to Katrina, New Orleans had been disassociated, to some extent, from its surrounding natural environment through urban development. However, after Katrina effectively brought the city's intricate connection to the surrounding water back into the daily lives of New Orleans residents and destroyed most of the manmade structures, the manmade social landscapes in the city were irreparably altered. Throughout the rest of this book, we will explore the ways in which nature's assertion of itself on the daily lives of New Orleans and Gulf Coast residents has altered the social and political landscapes of the area.

This work brings together the theoretical insights from Lefebvre, Giddens, and others. We argue that landscapes are superimposed and exist as layers simultaneously stacked atop of one another. However, our analysis synthesizes the different landscapes as they attempt to recover at various rates and return to a "predisaster" state. The center of our analysis rests on understanding the praxis of human interaction as survivors navigate the social, political, and economic landscapes with their rules and regulations that "exist only in and through the activities of human agents" (Giddens, 1989, p. 256). Critical to this study is the understanding of the "structuring properties [rules and resources] . . . which make it possible for discernibly similar social practices to exit across varying spans of time and space, which lend them to systematic form" (Giddens, 1984, p. 17) and result in resources needed to order social life within one of the landscapes. Miller and Rivera (2007a) contend the following:

> The disaster landscapes exist not only as a structure but also as a parallel to the "normal structure" (pre-disaster) as individuals constantly refer back to it as a reflexive notion of the way reality ought to be represented. Critical to the understanding of human agency in the aftermath of a disaster is the existence of the disaster social structures, such as the socio-cultural, socioeconomic, and the political landscapes. (p. 144)

The rules and regulations governing behavior are different in the new context. From the disaster experience emerges a new set of rules for governing life in the natural and human-constructed landscapes. Rules become redefined and internalized as the "new normal" so humans can navigate the physically and socially constructed landscapes simultaneously.

Notes

1. See also Lefebvre, 1991, p. 143.
2. "These bonds are developed through long-term, focused involvement in a residential setting. Through the purposeful and satisfying concentration of the multiple routines of daily life in a geographic location, the residential environ is . . . imbued with positive affect" (Feldman, 1990, p. 187-188).
3. This research was presented as an invited paper at the Seventeenth United Nations Regional Cartographic Conference for Asia and the Pacific, Bangkok, Thailand, September 18-22, 2006.
4. The Lake Okeechobee Hurricane caused 2,500 deaths, the Galveston Hurricane caused 8,000 deaths, and Hurricane Katrina caused 1,604 deaths.
5. Galveston is an island city located on the eastern end of Galveston Island about 30 miles long and from 1.5 to 3 miles in width between Galveston Bay and the Gulf of Mexico. The peak elevation of the island is only 8.7 feet above sea level (Heidorn, 2000).
6. Among the duties of Dr. Isaac Cline were Section Director of the Texas Section of the United States Weather Service, Executive Director of the Galveston Weather Station, and Executive Director of the Cotton Region Services for Texas (Heidorn, 2000).

7. An interesting fact about the recovery of the dead rests in the Galvestonians' attempt to have mass burials at sea. This attempt failed, resulting in large numbers of bodies washing ashore and covering the landscape.

Chapter 2
The Physical Landscape

A port or two would make us masters of the whole of this continent.
—René-Robert Cavelier, Sieur de La Salle, ca 1684 (Campanella, 2002, p. 15)

Unlike any other region in the United States, the Gulf Coast is anchored by the vast delta network of the Mississippi River, which strategically places New Orleans in a unique geographic position. Present day New Orleans is the first major city travelers encounter when leaving the Gulf of Mexico as they enter the North American mainland via the Mississippi River. So unmistakable is the river's influence that from this vast fertile plain, the North American continent becomes navigable; the history, culture, and geography of the region are dominated and shaped by it.

This chapter outlines the socio-political, economic, and historical contexts of land development in New Orleans and the surrounding environments. By analyzing the history of the region and the events during its colonial period, we draw parallels between land development policies and river management strategies; moreover, we discuss the impact of centuries-old practices on the topography that have subsequently led to catastrophic devastation to coastal and inland regions, the loss of life and property, and the destruction of the coastline. We contend that the devastation and reconstruction of the physical landscape of New Orleans serves as a metaphor for the entire recovery effort in the effected area.

The Evolution of New Orleans' Landscape

Hernando DeSoto was the first European to encounter the native peoples of this area. DeSoto's conquistadors swept through the land, seeking to take with them as many treasures as they could find. Later, the landscape of the lower Mississippi River was first charted by Europeans in the early spring of 1682 when the French expeditions of LaSalle reached the delta. They claimed the vast land drained by the Mississippi and all of its tributaries and named the land Louisiana in honor of King Louis and Queen Ana. Unlike DeSoto, LaSalle and subsequent French explorers sought to establish friendly relations with the native peoples through trade. The French attempted to treat the native people with dignity; Iberville later (as cited in Richard, 2003) wrote:

We all smoked an iron calumet [peace pipe] I had made in the shape of a ship with the white flag adorned with the flurs-de-lis and ornamented with glass beads. [I then gave them as a present] axes, knives, blankets, shirts, glass beads, and other things valued among them, making them understand that with this calumet I was uniting them to the French and that we were one [nation] from now on. (p. 6)

In 1698, Louis XIV chose two brothers from Montreal to lead an expedition to establish a colony in the land claimed by LaSalle and firmly assert France's claim to the land. This move to claim and chart land not only helped spread French mercantilism but also increased French wealth and political power among the other 18th century world powers. During the 18th century, in an effort to extend French power and influence, the crown's quest to expand the empire moved to the Americas and the mapmakers provided the crown with its imagery and authority to its claims[1] (Petto, 2007). Pierre Le Moyne, sieur d'Iberville and his younger brother Jean-Baptiste Le Moyne, sieur de Bienville reached the area and found it to be a landscape rich in Native American cultures that dated back several centuries before the birth of Christ (Roberts, 2003). A French colony at Biloxi, Mississippi was founded shortly thereafter in 1699 (Biloxi was a part of Louisiana at the time), and European settlement of contemporary Louisiana began (Congleton, 2006).

Expeditions were led to the region to seek an appropriate location able to support a population. One such expedition, led by de Bienville,[2] was successful in doing just that. On June 10, 1718, Bienville[3] recalled, "I myself went to the spot, to choose the best site" (Brinkley, 2006, p. 6). Jean-Baptiste Bénard de La Harpe recorded as follows:

In the month of March, 1718, the New Orleans establishment was begun. It is situated at 29°50' in flat and swampy ground. . . . The Company's project was, it seems, to build the town between the Mississippi and the St. John river [Bayou St. John], which empties into Lake Pontchartrain; the ground there is higher than on the banks of the Mississippi. This river is at a distance of one league from Bayou St. John, and the latter brook is a league and half from the Lake. A canal joining the Mississippi with the Lake has been planned which would be very useful even though this place served only as warehouse and the principal establishment was made at Natchez. The advantage of this port is that ships of [left bank] tons can easily reach it. (as cited in de Villiers du Terrage, 1920, p. 179)

Moreover, Pierre-Francois-Xavier de Charlevoix, a Jesuit priest who toured New Orleans in 1722, noted that the environmental conditions were treacherous:

Justest notion you can form of it is, to imagine to yourself two hundred persons . . . on the banks of a great river, thinking upon nothing but putting themselves under cover from the injuries of the weather. Even though Charlevoix worried for the colonists, he nonetheless envisioned a bright future for the city, writing that "Rome and Paris had not such considerable beginnings . . . and their founders met not with those advantages on the Seine and the Tiber, which we have found on the Mississippi, in comparison of which these two rivers are no more than brooks. As a result, he predicted that this wild and desert place, at present almost entirely covered over with canes and trees, shall one day . . . become the capital of a large and rich colony. (Charlevoix, 1923, as cited in Kelman, 2003, pp. 6-7)

Charlevoix, Bienville, and other early settlers commented on the proposed city's relationship with its surroundings and believed that the promise of New Orleans' situation outweighed the shortcomings of its topography. Henry Murray, a visitor to the city in the mid-19th century, summed up this viewpoint: "New Orleans is surprising evidence of what men will endure, when cheered by the hopes of an ever-flowing tide of all mighty dollars and cents" (Charlevoix, 1923, as cited in Kelman, 2003, pp. 6-7) (Map 2.1).

Southeast Louisiana was a strange place, and food was in short supply. Settlers tended not to live long after their arrival. After Iberville left Louisiana to return to France to champion the cause of the colony, Bienville was made governor until 1712. As conditions grew worse and Louisiana was on the verge of collapse, the French government decided to turn their Louisiana "problem" over to a private company that would sponsor the colony as an investment, control its development, and populate the land. Antonie Crozat became proprietor of the colony and appointed Antoine Laumet de La Mothe, Sieur de Cadillac as Governor General of Louisiana in 1710 while Bienville remained second in command. After a short time, Crozat ceded his rights to Louisiana back to the crown because he found his venture was a losing proposition.

The King took possession of the colony and found another investor, The Company of the West, which was headed by John Law, who published several descriptions of the colony based on Father Louis Hennepin's *A New Discovery in a Vast Country in America: A Voyage Narrative of the Upper Mississippi*.[4] This account, considered unreliable by some, sparked tension between the French and the British, who asserted theoretical claims that their original charters from their North American colonies extended indefinitely from the Atlantic shores (Crouse, 2001). With a description and the formation of the Louisiana Company, Law marketed Louisiana across Europe. He enticed investors and settlers to partake in a promising plan that "resolved to establish, thirty leagues up the river, a burg which should be called New Orleans [La Nouvelle Orléans], where landing would be possible from either the river or Lake Pontchartrain" (de Villiers du Terrage, 1920, p. 173). By 1717, advertisements continued to exalt the vast opportunities for exploiting the landscape in the Louisiana Colony. Speculation of the over-exaggerated claims led to the realization that a city still needed to be founded and, at that point, the site was nothing more than a swampy matrix. The city was finally founded approximately three years later once its advantages and disadvantages were compared with other potential sites.

Although the Louisiana Company also failed, Law's dealings yielded one important group that radically changed the landscape of the region: the German immigrants. Once in Louisiana, they were persuaded to live in the areas now known as the St. Charles and St. John the Baptist Parishes. They were able to work the land and transform it into something more suitable for agriculture (Roberts, 2003). Another group also changed the physical landscape. In 1719, at the recommendation of Bienville, two ships, the le Duc du Maine and the L'Aurore, landed in the colony carrying slaves from the Senegambia area of Western Africa. These slaves had extensive knowledge of farming, boat building, and metal work. It took them years to clear the heavy cypress and cane, but by the 1730s New Orleans was a major commercial center and a plantation society (Roberts, 2003). Bienville again resumed the role as French Colonial Governor, and in 1733, under his administration, the city of New Orleans was founded.

The Place: From Island to City

In the early 1700s, when Jean-Baptiste Le Moyne, sieur de Bienville, was led through the marsh and thickets, he and his Native American guides came across a crescent in the river. It was here, on the only solid soil he could find high enough to sustain a major port city, that he situated the city. Because it was virtually surrounded by water, the settlement founded was originally named L'Isle de la Nouvelle Orléans (the Island of New Orleans); today, it is known as New Orleans, or the Crescent City (de Villiers du Terrage, 1920; Kelman, 2003).

In 1718, the urban landscape began to take its early form when Bienville initiated a plan to build a permanent settlement. From 1721 to 1722, the region was surveyed and the city's design was developed by LeBlond de la Tour and Adrien de Pauger (Map 2.2). They used a symmetrical, grid-style configuration with a central location dedicated to the public institution. The first plan called for a sixty-six block grid, which was called the old city. They began by identifying lots keyed to owner assignments. According to Campanella (2002), "the remote colonial outpost remained within the neat plan until 1788, after a catastrophic conflagration and growing pressure for more space triggered New Orleans' first expansion beyond its original confines" (p. 92). During the Spanish rule, the burgeoning city was forced to spread beyond the protection of the higher elevated soil deposited closest to the river.

When the territory of Louisiana became a Spanish possession in 1762 by way of secret transaction between the French and Spanish governments, colonial leaders drove out the first Spanish Colonial Governor, Antonio de Ulloa, in 1768. He was replaced by Alejandro O'Reilly, who imposed Spanish law (Magill, 2003). Most notable of the city's landscape changes introduced by the Spanish were the increase in population and the increased urban expansion. The city's population reached 3,190 residents in 1769; 4,980 residents in 1785; 5,331 residents in 1788; and 8,056 residents in 1797 (Magill, 2003). As the population grew and the region prospered, the Spanish governors instituted municipal changes, including the building of the Corondelet Canal (the extension from Basin Street to Bayou St. John) to facilitate drainage and navigation. During this same time, two major fires, one on March 21, 1788, and another in December 1794, destroyed large parts of the city. With the city in ruins from the fires, the Spanish rebuilt the city on a much grander scale, with larger homes, more stringent building codes requiring fire walls to be constructed, and a reliance on more brick and cemented timbers (Magill, 2003). Although the city began to reach its original boundaries, the river remained intimately connected to the city's development, firmly rooting its resident's sense of place.

At first, the planned development followed the well-established flood zones; however, as residents sought land beyond the city with the creation of faubourgs,[5] they began to encounter seasonal flooding, which promoted the construction of levees and the employment of techniques to cope with the constantly shifting river that shaped and continued to change the dynamic urban-riparian environment and physical landscape (Kelman, 2003). As long as people could control floods, they could do business in the region. But the residents and businessmen learned too late that the landscape, which ultimately gives shape and form to the dynamic floods, depends on the floods for its economic vitality.

In 1812, Louisiana became the 18th state of the United States, and the push was made to change the topography of the city by extracting the boundaries and digging more canals for navigability in an attempt to get the goods produced there to markets in the East and Midwest (Map 2.3). By 1836, the city was formally divided into three self-governing "municipalities": the French Quarter (the original city), St. Mary (the "New American city" in Faubourg St. Mary), and downtown (the modern day 9th Ward) (Lewis, 2003). Since 1718, New Orleans had been economically obligated to streamline this linkage (a passage route that would be quick and easy) to allow for the swift passage of vessels. Although for years many local merchants had maintained a reverence for the Mississippi River as the transportation link, complacency, termed "the Mississippi Obsession" by one historian, led to intense competition from northern railroads in the late 1800s (Clark, 1967). Much of the expansion included navigable canals, such as the New Basin Canal, which connected Lake Pontchartrain to the back part of the city by 1838 (Maps 2.4-2.6).

Although the expansion of the city followed the economic expansion of the elite Creoles, who were born and lived in the French territory of Louisiana, the American elite also changed the geography of the city by expanding their economic interests and settlement patterns uptown into what is now called the Garden District. As Americans moved into the Ohio Valley and the Louisiana Territory in the 19th century, commerce along the Mississippi River expanded and New Orleans became a small city. River commerce also accelerated when new steamboats came into service, allowing goods to move upriver nearly as easily as down. Due to these changes, the low-lying trading post at New Orleans became a major port city. By 1860, it included nearly 170,000 residents and was one of the largest cities in the country (McNabb & Madère, 1983). Growth continued for the next century as river traffic continued to expand and valuable reserves of oil and natural gas were discovered along the Gulf's continental shelf. In 1960, New Orleans' population reached 627,525 (Congleton, 2006, pp. 8-9). According to Campanella (2002), "a city [was] formed, restricted to the higher lands of natural levees and dramatically reflecting topography in its shape and growth" (p. 80), but the threat of flooding haunted the city from the beginning.

A History of Water, Landscape, and Development

Since the beginning of the city, the river, which gave the city a sense of place and rooted its residents to that place, often invaded the living quarters of settlers.

The city's founder stated, "In April 1719, when New Orleans was a year old ... Bienville reported that the Mississippi floods were regularly drowning the settlement under half a foot of water. [Furthermore], he suggested building levees and drainage canals as soon as such work was required of land owners" (Magill, 2005-2006, p. 33). It is the river's influence that fundamentally created the underlying terrain. The terrain emerged from an alluvial valley and built a deltaic plain. The ebb and flow of the river repeated itself over thousands of years in the way of routine flooding. Eighteenth century settlers and developers worked with the land to plan the city along its natural boundaries.

Since the 18th century, when French colonial administrators required land claimants to establish ownership by building levees[6] along bayous, streams, and rivers, people have tried to dominate the region's landscape and the forces of nature (Dean & Reukin, 2005). As early as 1719, the first manmade levees were built. By 1735, the levees lined the river from Englishturn to present-day Reserve, to the Old River region by 1812, and to Greenville, Mississippi by 1844 (Campanella, 2006a; Walker & Derro, 1990), but the seasonal flooding was always a threat to the region.

> Levees have mostly held back the Mississippi, but occasionally crevasses have sent water coursing into New Orleans. On May 5, 1816, a levee gave way at the McCarty plantation in present-day Carrollton and within days water was filling the back portion of the city. Water reached St. Charles Avenue, extended to Canal and Decatur Streets, flooded the French Quarter and [in] Faubourg Marigny to Dauphine Street. Only the highest parts of the natural levee remained dry. (Magill, 2005-2006, p. 35)

Furthermore, Magill (2005-2006) contends the following:

> On May 4, 1849, the Mississippi broke through the levee at the Sauvé plantation at River Ridge. Within four days, water reached the New Basin Canal, a week later it crossed St. Charles at Napoleon Streets, and ten days after that, [it reached] Canal and Dauphine Streets with Barone Street flooded. The French Quarter had water to Bienville and Dauphine Streets. (p. 35)

Between 1840 and 1860, a series of harsh floods[7] made it apparent that the flooding problem was too large to be handled by local governments. The issue was critical to national interests. Much of the unprotected and unoccupied land was in the public domain. If reclaimed and protected against overflow, the land would become a national economic asset.

Two severe floods in 1849 and 1850 convinced the federal government to provide financial assistance to construct a continuous levee system. After multiple instances of major flooding, the federal government recognized the problems associated with such periodic flooding and enacted the Swamp Land Act of 1849, which applied only to Louisiana. In contrast, the Swamp Act of 1850 extended to other states (Davis, 2000).

Not only does the city have a history of river floods, but documents of tropical storm and hurricane seasons also record a long history of destruction. Since the settlers began documenting events, hurricanes have been recorded.

In 1778, 1779, 1780, and 1794, hurricanes struck the region destroying buildings and sinking ships. Storm surges swamped areas south of New Orleans. The worst storm of the period was "The Great Louisiana Hurricane" of August 9, 1812. It topped barrier islands and drowned Plaquemines and St. Bernard parishes and the region surrounding Barataria Bay under 15 feet of water. . . . In 1831, The Great Barbados Hurricane tore through the Caribbean and struck west of New Orleans. The area south of town once again was inundated by storm surge, while in the city 3 feet of water flooded Lake Pontchartrain's south shore. The Mississippi levee at St. Louis Street broke sending water into French Quarter streets. Heavy rainfall submerged the back portion of town, forcing residents to flee. (Magill, 2005-2006, p. 34)

Because New Orleans' highest elevations are along its natural levels, the silt deposited embankments rise more than 10 feet above sea level in some places; however, much of the city rests below sea level, creating the bowl effect between the Mississippi River and Lake Pontchartrain due to its naturally occurring ridges. The first of such ridges, the Metarie Ridge (the site of a former bayou), stretches from Harahan along Metarie Road near Bayou St. John where it meets the Gentilly Ridge (another old bayou), which traces the Chef Menteur Highway (Magill, 2005-2006). The Esplanade Ridge runs perpendicular to the linear crests of the Metarie Gentilly Ridges to follow what is now Bayou Road. These ridges rise approximately four feet above sea level, and their basins drop to nearly eight feet below sea level.

Citizens sought land for settlement beyond the city's walls, but a vast cypress swamp grew north of the city and stretched to Lake Pontchartrain, where the basin of the swamp dipped to more than 6 feet below sea level only to rise again to meet sea level at the shores of Lake Pontchartrain. The threat of river flooding and storm surges was combated by canal and levee construction as development continued. This allowed the city to expand and meet the needs of the growing American economy. Water trapped within the new boundaries was drained and pumped over the lower lying areas into the lowlands as residents brought in earthen landfill from the newly excavated canals to stimulate commerce. Campanella (2002) further contends that "as a result, many of New Orleans' topographical problems—the challenges that restrained and threatened the city for centuries and made residents curse Bienville's stubborn attachment to the site—have been conquered" (p. 80). As a result of the crisscross patchwork system of the manmade canals,[8] residents could live with some degree of protection from the routine floods. However, major floods continued, with notable ones occurring in 1850, 1858, 1862, 1865, and 1874. In 1879, after this history of "local" flood disasters, the United States Congress created a new agency that would hold authority over the entire river system—the Mississippi River Commission. Under the authority of the Mississippi River Commission, the Army Corps of Engineers took charge of flood control and transport within the entire basin. Some risk of flooding on the river system always remains, but river flooding in Louisiana declined after 1927, largely due to more sophisticated efforts at water-level management (Congleton, 2006).

As noted earlier, flooding and the need for a massive levee system began in the 18th century. By 1812, the settlers of the Louisiana Territory had constructed levees from the east bank of the Mississippi River to Baton Rouge, 130 miles upstream, and on the west bank as far as Point Coupée; 165 miles of levee construction was done upstream as a measure to tame the river (Tibbetts, 2006). City leaders in New Orleans viewed the draining of swampland north of New Orleans not only as a way to positively affect economic development but also as a way to curtail some of the health effects of yellow fever and mosquitoes. By the 20th century, New Orleans had a well integrated public works department in place that was responsible for draining the swamps and wetlands. "It was the drainage of the lower areas that allowed for suburbanization" (Tibbetts, 2006, p. A42).

The land in and closest to the original city was the most desirable for urban expansion. Before clearing the swamp and marshy areas, the city's land growth was bound by its topographical tie to naturally occurring high ridges beyond the river's banks, such as the Esplanade, Metairie, and Gentilly Ridges. Prior to the 19th century, habitation of these areas directly adjacent to the outskirts of New Orleans was mostly composed of raised fishing camps and squatter shanties. Urban development (and we add culture, politics, and economy) was so intricately tied to urban topography that the New Orleans city maps of the 19th century resemble elevation maps (Campanella, 2006a). The convenience, accessibility, and well-drained nature of some portions of the land made it a candidate for expansion and further subdivision that characterized the development of the municipality beyond its original boundaries.

> As Colten observes, the draining, leveeing, and development of Lake Pontchartrain area wetlands altered what had been a wide marsh and swamp buffer between the city and the lake, placing new residential districts in danger and removing the ability of tracts to the north of the urban core to protect it from waves and storm surges flowing off the lake. (Colten, 2005, as cited in Brookings Institution, 2005, p. 26)

> In 1835, Ingraham noted: "I have termed New Orleans the Crescent City . . . [because it was] built around the segment of a circle formed by a graceful curve of the river." (Ingraham, 1935, as cited in Campanella, 2006a, p. 94)

As described, the topography drove the city's development, the local and national ideology, and the relationship between nature and human development. The 2005 Atlantic hurricane season was the most active season on record and produced three storms that reached category 5 intensity: Katrina, Rita, and Wilma. Katrina was not the strongest of the three storms in terms of wind speeds or central pressures, but converging factors—primarily its strength and landfall location along the Gulf Coast—made it the most devastating and costly hurricane in the history of the United States. It is estimated that Katrina impacted 90,000 square miles (an area nearly the size of the United Kingdom), displaced more than 1 million people, and killed more than 1,300 people (Federal Emergency Management Agency, 2005). Hurricane Katrina exposed the coastal populations of Louisiana and Mississippi to an unprecedented combination of natural forces and human failures (Cutter et al., 2006).

When Katrina came ashore, it caused major changes to the landscape. Even though the present-day landscape was not completely created by Katrina, in the eyes of some, it is the result of a variety of political, economic, and land management decisions that rendered an environmentally sensitive region more susceptible to abrupt environmental devastation. According to Pabis (2000), nearsighted economic interests motivated planters to build levees hastily, blocking outlets and draining swamps without concern for the long-term effects of the physical alteration of the landscape. By confining the river to a single channel, it forced sediment that had once accumulated on alluvial lands into the main channel to eventually settle along the riverbed, thereby raising the level of floods. Eventually, the Mississippi would overwhelm the levee system, and New Orleans would be under several feet of water.

Prior to Katrina, scientists estimated the loss of Louisiana coastal wetlands at between 25 and 35 miles per year, or the equivalent of a football field every half hour (Bowser, 2006). Long-term protection, preservation, and restoration of Louisiana's wetland resource base cannot be accomplished without diverting sediment-laden water from the Mississippi River.

> The crevasses and overbank flooding have historically renourished the wetlands. At present, crevasses and overbank flooding are no longer "nourishing" the wetlands. Humankind engineered a levee system to protect against flooding. These levees successfully deprived the wetlands of valuable sediments, which directly affects the natural resource base, local communities, and public infrastructure—impacts that are real, undeniable, and to some unimaginable but deserve real solutions. From a historical perspective, crevasses are sediment conduits; they may be one element in the solution to wetland loss. (Davis, 2000, p. 106)

These coastline changes result from decades of alterations as the silt and mineral deposits, which once worked to build the lower delta region, were channeled and redirected into the waters of the Gulf of Mexico. The most dramatic evidence of such changes can be seen in the satellite images of the area before and after Katrina. Areas that were once land masses and covered with greenery currently lie underwater. When comparing the coastal maps of 1839 with the current topography, one can see an extensive network of solid soil that once served as a natural buffer for the region against tropical storms and hurricanes. However, with the passage of time, the channeling and dredging of the Mississippi River and the construction of the extensive levee systems have changed the spatial configuration and population density of the city (Bowser, 2006).

Currently, the water from the Gulf of Mexico is closer to heavily populated regions. Randy Hanchey of the Louisiana Department of Natural Resources warned: "I don't know that the Gulf is going to show up on the doorstep of New Orleans in 50 years, but you're going to have shallow, open water all the way to New Orleans metropolitan area if we don't do something about restoring some of these wetlands" (as cited in Bowser, 2006). The giant marshlands to the south of New Orleans that occupy a large portion of Southeast Louisiana, Mississippi, and parts of Alabama were once helpful in breaking down the storms as they approached more populated inland areas. The coastal vegetation buffer zone is essential to increasing or decreasing the region's susceptibility to major storm surges. For instance, researchers found that costal vegetation made a dramatic difference during the 2004 South Asian tsunami; shorelines lined with mangrove forests suffered significantly less damage than areas where the tidal wave met land denuded by human activity. According to Tidwell (2006), "scientific analysis conducted soon after the disaster suggests that just 301 trees per 120 square yards in a 100-yard-wide belt could diminish the maximum tsunami impact by more than *90 percent*" (p. 25).

In Louisiana, however, this land is disappearing. The land along the Gulf Coast is constantly earmarked for development initiatives. Moreover, as much as 80 percent of the nation's coastal wetlands loss occurred in Louisiana between 1932 to 2000; the state lost 1,900 square miles of land that are now under the waters of the Gulf of Mexico[9] (Tibbetts, 2006). Tibbetts (2006) argues that if nothing is done to stop this process, by the year 2050, the state will lose another 700 square miles of marshland and one-third of the documented 1930s coastline will have vanished. Most importantly, "New Orleans and [the] surrounding areas will become even more vulnerable to future storms" (Tibbetts, 2006, p. A40). For example, "when Teddy Roosevelt was president, Barataria Bay was more grass than water, and the marshy peninsula of Plaquemines Parish was three times wider than it is now. In Breton Sound, the combination of grasses and barrier islands created a muscular bulwark of land" (Tidwell, 2006, p. 25). Such land formations are not there today; Katrina was able to simply pass over mostly open water as it barreled toward New Orleans.

Abrupt Landscape Change

On August 25, 2005, a hurricane watch was issued as the winds of the tropical depression that would become known as Hurricane Katrina reached forty-five miles per hour. Fed by the warm, summer waters of the Caribbean, the storm grew rapidly, its winds intensified to over seventy miles per hour, and Florida was put under a hurricane watch. Katrina reached the Florida coast just north of Miami Beach with category 1 strength and a storm surge between two and four feet. As Katrina crossed the Florida Panhandle, she weakened slightly but reentered the Gulf of Mexico where the waters refueled the storm system, causing Katrina to regain strength. Within a matter of days, the storm reached category 4 status, with sustained winds in excess of 145 miles per hour. Within 72 hours of crossing the Florida Panhandle, Katrina's winds reached 175 miles per hour, the storm was over 10,000 miles wide, and the surge was as high as 28 feet. She had become the storm of cataclysmic proportions that would wreak havoc and unimaginable destruction to the coastlines and landscapes of Louisiana and Mississippi, both of which were directly in Katrina's path (CNN Reports, 2005). Compared with storms such as Ivan, Camille, and Betsy, "This hurricane can cause not just trees to be blown down and mobile homes and things like that. . . . This storm . . . can cause some catastrophic structural damage from not just homes along the shoreline, but homes inland"[10] (CNN Reports, 2005, p. 10). Whereas others stated as follows:

> This is one of the worst-case scenarios. It's kind of a doomsday scenario. Very rarely have we ever seen nature conjure a storm this powerful in the past 100

years. It ranks up there with storms such as Camille and Andrew and Galveston. [There] is no way to soften the blow of this storm whatsoever. It's packing the worst possible energy in terms of ocean atmosphere, and all unimaginable energy. (Dr. Jeffrey Haverston, NASA, as quoted in CNN Reports, 2005, p. 10)

As the evacuation of the city began, city officials realized that the landscape would be changed:

> The problem we have with this storm, [is] if those storm surges are that high, they will top our levels and there will be lots of water in the city of New Orleans. (Mayor Ray Nagin, 2005, as quoted in CNN Reports, 2005, p. 10)

By 10:00 a.m. on Sunday, August 28, 2005, Mayor Nagin ordered a mandatory evacuation of the city after the following National Weather Service Advisory:

> ... DEVASTATING DAMAGE EXPECTED ...
>
> HURRICANE KATRINA ... A MOST POWERFUL HURRICANE WITH UNPRECEDENTED STRENGTH ...
>
> MOST OF THE AREA WILL BE UNINHABITABLE FOR WEEKS ... PERHAPS LONGER. AT LEAST ONE HALF OF WELL CONSTRUCTED HOMES WILL HAVE ROOF AND WALL FAILURE. ALL GABLED ROOFS WILL FAIL ... LEAVING THOSE HOMES SEVERELY DAMAGED OR DESTRYOED.
>
> POWER OUTAGES WILL LAST FOR WEEKS. WATER SHORTAGES WILL MAKE HUMAN SUFFERING INCREDIBLE BY MODERN STANDARDS.
>
> (National Weather Service, 2005)

At 6:10 a.m. on August 29, 2005, Katrina made its first landfall near Grand Isle, Louisiana, as a Category 4 storm with a cloud span of over 1,300 miles and winds affecting places over 400 miles away from the center (CNN Reports, 2005). As the streets of New Orleans filled with water, the conditions deteriorated rapidly and the landscape became a hydroscape. High winds downed power lines, and flying projectiles made it unsafe for humans and animals to be outside. Even areas outside the city had "winds of over 90 to 180 miles per hour [and were] not fit for humans" to be in (Rob Marcians as quoted in CNN Reports, 2005, p. 20). And just as quickly as the water rose in the streets of New Orleans so too did the water rise in the homes of the residents who remained in the city. With nowhere to run, chaos began to unfold in the streets and water-filled homes; residents were forced to retreat to their attics where they remained for days, in many cases with no provisions, until rescue efforts began.

As residents made their way into the destruction, a new world emerged; it was as if the landscape that gave birth to the culture that defined the Mississippi delta, particularly New Orleans, had been under siege by the most brutal force of nature the region had ever seen. Whole neighborhoods lay in waste as the flooding overtook all in its path. The force of the water was so great that a barge broke through the levee and floated through parts of the Lower 9th Ward.

The city of New Orleans did not sustain a direct hit from Katrina, but suffered from the damages caused by the raging water and wind and the flooding that remained in some areas of the city for weeks. By August 30, 2005, nearly every inch of soil between the French Quarter and St. Bernard Parish was under water and nearly 40,000 homes were submerged. Reports confirm that the first levee breach was in the 17th Street Canal Levee, which held back Lake Pontchartrain. As millions of gallons of water poured into the city, the breaches grew wider while more breaches were reported along the Industrial Canal. The water from these breaches inundated the Lower 9th Ward, where the water reached between 12 and 15 feet high in some places (CNN Reports, 2005, p. 32). In early May of 2005, one journalist predicted that:

> Soon the geographical "bowl" of the Crescent City would fill up with the waters of the lake, leaving those unable to evacuate with little option but to cluster on rooftops—terrain they would have to share with hungry rats, fire ants, snakes, and perhaps alligators. The water itself would become a festering stew of sewage, gasoline, refinery chemicals, and debris. (Mooney, 2005, as cited in Dyson, 2006, pp. 79-80)

During Hurricane Katrina, the storm surge from Lake Pontchartrain was not strong enough to top the 17th Street or London Avenue canals, which ultimately flooded large parts of the city following their structural collapse. According to Magill (2006), if the floodwalls had just been overtopped, flooding would have lasted only a few hours—the length of time for the storm to pass—instead of the several days it did. Forensic investigations revealed that the floodwalls broke because they were poorly designed, had used obsolete elevation data, and were set on top of weak levee soils. Along with these failures, the man-made waterways—the Industrial Canal, the Intracoastal Waterway, and the Mississippi River Gulf Outlet—channeled the storm surge directly into the urban areas of eastern New Orleans and St. Bernard Parishes (Magill, 2006).

Not only did the water rise, but a chemical sheen, known as toxic gumbo, could also be seen on the top of the water. By the end of the third day after Katrina, New Orleans became "a unique blend of chemical waste, gasoline, and the inevitable coffins from the city's above-ground cemeteries: toxic gumbo it's called. . . . [Furthermore,] by evening, with most authorities involved in the search and rescue efforts, looting became widespread in all parts of the city" (CNN Reports, 2005, p. 35). Later, the landscape was described by the *Washington Post* as a place where "Houses walk and the dead rise up" (Roig-Franzia, 2005, p. C1).

During this time, national television news viewers were riveted by stories of beatings, rapes, and murders in both the Superdome and Convention Center. It was later determined that even as conditions deteriorated, these crimes were largely sensationalized rumors (Magill, 2006). At one point it was reported that floodwater would reach 9 feet on St. Charles Avenue and 12.5 feet at the Mississippi River. These depths would have been impossible because the riverfront is several feet higher than St. Charles. Concurrently, it was reported that Lake Pontchartrain was approximately 3 feet above sea level, leaving elevated sections of the city dry, such as the Mississippi riverfront and large portions of Uptown (Magill, 2006). Citizens leaving their homes were forced to pass the hazards of downed power lines, scared animals, rusty nails, collapsing housing structures, and floating corpses. Many others were forced to leave relatives behind. Tens of thousands of New Orleanians were trapped in houses, hospitals, and hotels. They were also in the Louisiana Superdome and the Morial Convention Center ("shelters of last resort") where they desperately awaited food, water, and rescue (Magill, 2006, p. 49). The destruction of the landscape was so complete that the storm surge from Katrina leveled entire parts of the city, leaving the foundations of homes and little else.

Place and Landscape Change

The waters of the Gulf of Mexico were over 80 degrees Fahrenheit in some areas, adding the fuel of thermal energy to the monstrous storm. Most ominous was the storm's sheer size; it was 450 miles wide when it slammed into the Gulf Coast and left a swath of destruction from Moyan City, Louisiana, to Apalachicola Bay, Florida (*Times-Picayune* Staff, 2006). The storm first began to wreak havoc on the landscape with its high winds. Telephone poles and trees were the first casualties of the storm. Williams (2006) noted:

> Pine trees blow to the ferocious winds until the trees snap like twigs in a child's hand. One breaks several feet from its base, then another, then dozens, like popcorn beginning to pop on a kitchen stove . . . we wait for the tree that will smash the house—and us. (p. 27)

Then, the storm surge up-swelled, rolling across the vanishing wetlands in Southeastern Louisiana and filling the local canals and lakes. The overtopped levees and swelled bodies of water funneled water into the city and worsened the storm surge by channeling the water deep into the low-lying areas, which ultimately covered over three-quarters of the city.

The destruction caused by Hurricane Katrina was equated to the level of mass destruction seen after the Great Haushin earthquake (Kobe, Japan), the Indian Ocean tsunami, and even an atomic bomb. Scientists estimated that "Each hour, Katrina unleashed energy equivalent in its fury to five atomic bombs of the size dropped on Hiroshima" (*Times-Picayune* Staff, 2006, p. 28). The landscape, with debris littered where houses once stood, looked as if the atomic bomb had been dropped. The north shore of Slidell, Louisiana, was in the direct path of the storm and was destroyed. Boards and bricks were scattered in the streets, making them hazardous and impassable in many areas. The once prosperous marinas on the shores of Lake Pontchartrain lay in utter waste. As Hurricane Katrina eroded more than 100 miles of Louisiana's coastal marsh (more than what would normally be lost in a year), it further reduced the resiliency of the area to survive such an assault. With this buffer now reduced farther (*Times-Picayune* Staff, 2006), the area was vulnerable to the rampant destruction and ecological havoc strewn across the Gulf Coast.

Even before the levees broke, neighborhoods to the east became inundated. According to the *Times-Picayune* (2005), scientists reconstructed the events of the flood using land elevation data, interviews, and eye-witness accounts. As the storm approached the coast, its easterly winds from its northern quadrant pumped the rising surge into the marshes of Lake Borgne, an area east of St. Bernard Parish. There, two hurricane levees formed a V-shape that funneled water from an estimated 20-foot storm surge into New Orleans. "The surge reached the Industrial Canal before dawn and quickly overflowed both sides," (*Times-Picayune*, 2005) as reported by the canal loch master. At some point following the overflowing of the Industrial Canal and subsequent breaches in the structure, corps officials believe a barge broke loose and crashed through the floodwall, opening a large breach in the floodwall and accelerating the flooding of the Lower 9th Ward and St. Bernard Parish (*Times-Picayune*, 2005).

During this time, other levee breaches occurred, such as the one at the London Avenue Canal, and thousands of citizens found themselves trapped in water. The 17th Street Canal breach was confirmed publicly at 2:00 p.m. on Monday, August 29th, at which time approximately 4 feet of water was confirmed in one Lakeview neighborhood. As reported by the *Times-Picayune* (2005), at 3:00 p.m. the water reached knee-deep level under the Jefferson Davies Overpass, and the Interstate 10 dip under the railroad overpass lay 15 feet underwater; by late afternoon, people stranded on Interstate 10 near the Industrial Canal were able to see residents of the Lower 9th Ward stranded on their rooftops.

Flood waters rendered once dry land in the shadow of the protection of floodwalls into a lake with houses as far as the eye could see. "It would be Thursday—40 billion gallons later—before the water levels withdrew and outside the levees had been equalized" (*Times-Picayune* Staff, 2006, p. 37). With most of the city plunged into a grayish, blackish, murky body of water that seemed as if it would never recede, more ominous threats began to unfold as a landscape of risk.

Landscape of Risk

As television images of the horrific events gave the nation a hellish glimpse of the conditions during and immediately after the disaster, another story was unfolding. As many residents returned[11] weeks later to find a landscape full of risks and the damage inflicted by the flooding affecting all manners of property:

> Thousands of homes [were] devastated by the water; walls imploded, thick layers of sediment [were] strewn inside and out, mold coated the interior of the homes, roofs collapsed, and some homes [were] picked up by the water and moved into the middle of streets. Thousands of vehicles—cars, trucks, buses—were completely or largely submerged and destroyed and [were] sprinkled in unexpected locations across the area, many left precariously leaning up against roofs, teetering on fences, stacked on top of each other, or crushed by the powerful water. Entire commercial strips were completely destroyed, schools and public buildings [were] wrecked, power lines [were] toppled, including toxic-laden transformers, and gas stations [were] ruined. (Olson, 2005, p. 2)

As elected officials called for the speedy repopulation of the city and federal dollars began to pour into the local economy, potential environmental hazards swelled in silence and greeted the families as they returned. With petroleum sheens covering the floodwaters, *E. coli* bacteria ran rampant and industrial toxins covered parts of neighborhoods. The news media quickly labeled the standing floodwaters a toxic gumbo. Later testing of the storm waters found fewer elevated levels of contaminants than feared, but sampling was limited, and the water may yet present long-term problems (Manuel, 2006). These concerns prompted a team of scientists, led by John Pardue, director of the Louisiana Water Resources Research Institute at Louisiana State University, to conduct its own study of the New Orleans floodwaters. The report, published on November 15, 2005, in *Environmental Science & Technology*, stated categorically that, contrary to claims in the media, the floodwater was not a "toxic soup."

Frankly, risks existed in the region long before Katrina—400 sites scattered throughout the affected areas were identified as possibly needing cleanup because of their potential effects on human health (Manuel, 2006). Residents of the region, particularly those from Cancer Alley,[12] had grown accustomed to the barrage of contaminants that were a part of the petrochemical industry that extended as far north as Baton Rouge and as far south as St. Bernard Parish just below the city.

A study completed by Manuel (2006) at Louisiana State University found that the chemical oxygen demand and the fecal coliform bacteria levels were elevated in surface floodwater but were typical of normal stormwater runoff in the region. The report also claimed that lead, arsenic, and in some cases, chromium exceeded some of the drinking water standards. The study also found low concentrations (less than 1 percent) of benzene, toluene, and ethylbenzene, even in places where there was visible oil sheen. The report concluded that these data suggest that the Katrina floodwaters are similar to normal stormwater runoff, with some elevated lead and volitile organic compounds concentrations near the surface. However, according to Manuel, the study was limited to two areas within the city of New Orleans, and he warned that conditions could be different elsewhere, particularly in Lake Pontchartrain where floodwaters were being pumped.

The damages inflicted by Katrina go beyond the temporary changes in the landscape but express the long-term changes to the physical landscape of New Orleans. The farther inland Katrina went, the less severe the flooding, wind, and storm surge damage; however, this is not to say that there was not considerable damage done to property far inland. The cityscape lay as a contemporary Atlantis for more than two weeks, while crews tried to pump the water out of the city as quickly as possible.

In the two and a half weeks that had passed since Hurricane Katrina flooded the city, pumps had been working nonstop to return the water to Lake Pontchartrain. As portable pumps were brought in to supplement the permanent pumps already hard at work, as much as 380 cubic meters (380,000 liters or 23,190,000 cubic inches) of water were being pumped out of New Orleans every second, according to the U.S. Army Corps of Engineers (National Aeronautics and Space Administration, 2005).

Although the floodwaters did recede with human intervention, if the city was left without human and technologic intervention, the urban physical landscape of New Orleans may have been permanently changed, exposing those returning home to greater risks. In other places along the Gulf Coast, such as in Mississippi, the coastal regions were wiped clean of foliage, the beaches were pushed back, and vital port structures and residential areas were destroyed. Construction zoning will most likely change in response to the physical changes in the landscape so that future events have a less probable ability to destroy so much property, which will in turn affect resettlement patterns and the economics of the region.

The unknown risks associated with returning and clean-up were more pronounced:

> Aside from standing floodwater, hazards included a lack of potable water, sewage treatment, and electricity; chemical spills; swarms of insects (with anecdotal accounts of vermin and hungry domestic dogs); food contamination; disrupted transportation; mountains of debris; buildings damaged and destroyed; rampant mold growth; tainted fish and shellfish populations; and many potential sources of hazardous waste. (Manuel, 2006, p. 33)

Moreover, inside the homes, the "loss of power meant lift stations (which pump sewage uphill) could not work, causing sewage to overflow into houses and streets" (Manuel, 2006, p. 34).

Together, the United States Environmental Protection Agency (EPA) and the Louisiana Department of Environmental Quality sought to analyze floodwater, testing it for more than 100 hazardous pollutants including the following:

> Volatile and semivolatile organic compounds, metals, pesticides, herbicides, and polychlorinated biphenyls. They also tested for biological agents such as *Escherichia coli*. Their testing revealed "greatly elevated" levels of *E. coli*, as much as ten times higher than EPA's recommended levels for contact. According to the EPA, agency scientists found levels of lead and arsenic at some sites in excess of drinking water standards—a potential threat given the possibility of hand-to-mouth exposure. . . . Shortly after the hurricane struck, the U.S. Coast Guard began working with the EPA, the Louisiana state government, and private industries to identify and recover spilled oil along the coast. The team identified 6 major, 4 medium, and 134 minor spills totaling 8 million gallons. One of the most notorious spills occurred at the Murphy Oil Company plant, which dumped more than 25,000 barrels of oil into the streets of Chalmette and Meraux, Louisiana. As of December 7, the Coast Guard reported the recovery of 3.8 million gallons, with another 1.7 million evaporated, 2.4 million dispersed, and 100,000 onshore. (Manuel, 2006, p. 35)

The vast amount of data became overwhelming to the point that Kevin Stephens, director of the New Orleans Department of Health, stated as follows:

> I struggled every day to determine what (the data) meant and what to tell our health workers and the public. . . . What does "not an immediate health hazard" mean when you have people wading through the water. What does "not in excess of drinking water standards" mean? Is it a danger if you get your hands wet and touch your mouth? (Manuel, 2006, p. 35)

Journalists claimed the floodwaters were a toxic gumbo of dangerous chemicals and microbes, raising fears that any contact was a health risk. As returning citizens received permission to enter some neighborhoods but not others, anxiety grew daily regarding the long-term health effects of the local environment, and additional concerns about the water persisted. Not only was it a concern due to its potential toxins, but it also served as a place where mosquitoes and other insects could breed, posing a host of other environmental health concerns.

The force of the storm surge not only brought with it a variety of environmental health concerns, but it also destroyed, and in some cases physically rearranged, the existing built environment. So complete was the property destruction in some areas that not a single house remained. In other communities, houses that were not destroyed were shifted from their foundations several feet across the street, making the streets impassable, or were, in some cases, transported into new neighborhoods. Rusty nails and asbestos from older buildings littered the streets as if they were common garbage. Such hazardous debris remained in some areas for a year before residents were permitted to return. Those hearty enough to brave the oppressive heat, downed power lines, and impassable roads were greeted by the smell of the masses of dead and rotting fish lying in the street and slippery mud and muck that seemed to only get deeper the farther they went. The look, feel, and smell of death were all around, and all was still and quiet. As with their houses, residents' personal effects were relocated. In some instances, boats from as far away as Lake Pontchartrain lay atop the homes in the Lower 9th Ward.

As soon as survivors were allowed to return and able to face the strange, newly arranged topography that was their home or community, they only found more anguish. Not only was the task of clearing and rebuilding ahead of them, but there were also the health dangers of growing mold and fungus. Fast-spreading mold spores grew in the sweltering heat to the point that many of the mold-covered homes posed health hazards, resulting in the need to be "gutted" or torn down.

Filamentous micro fungi (mold) can threaten human health by releasing spores that become airborne and are inhaled, causing illness. Some molds produce metabolites (mycotoxins) that can initiate a toxic response in humans or other vertebrates (Robbins, Swensonn, Nealley, Gots, & Kelman, 2000). Repeated exposure to significant quantities of fungal material can result in respiratory irritation or allergic sensitization in some individuals (Bush, Portnoy, Saxon, Terr, & Wood, 2006; Solomon, Hjelmroos-Koski, Rotkin-Ellman, & Hammond, 2006). Sensitive individuals may subsequently respond to much lower concentrations of airborne fungal materials. Of the thousands of types of fungal spores found in indoor and outdoor environments, adverse health effects in humans have most frequently been associated with alternaria, aspergillus, cladosporium, penicillium, and stachybotrys (Hossain, Ahmed, & Ghannoum, 2004; Jarvis & Miller, 2005; O'Driscoll, Hopkinson, & Denning, 2005; Solomon et al., 2006; Stark, Burge, Ryan, Milton, & Gold, 2003).

According to the Associated Press (2005), the mold formed an interior landscape version of the kudzu, a non-native vine species that once introduced to the southeastern United States grew to cover the landscape and strangle all other plants in its path. "A survey by the Centers for Disease Control and Prevention in late October found that 46 percent of randomly selected homes in the New Orleans area had visible mold growth, and 17 percent had heavy mold coverage" (Centers for Disease Control and Prevention, 2006). For some homes lying in 7 feet or more of water for weeks, mold grew in large black and white colonies and covered everything. Many feared that the problem would continue as the mold dried and released its spores to form new colonies in parts of the home that had not been touched by the water or that had already been treated with a 10 percent bleach and 90 percent water solution. Homeowners donned protective goggles, gloves, galoshes, and environmental masks to enter their homes and begin the rebuilding process.

Any structure that was in the path of the hurricane was at risk of being its next victim, and the cemeteries were no exception. In a report issued on January 22, 2006, in the *Times-Picayune*, nearly 6 months after the storm, Brown (2006) reported that "in St. Bernard Parish, 80 tombs were broken open or washed away at a single graveyard, The Merrit Cemetery in Violet, during Hurricane Katrina" (p. A:1). Brown further noted that "in Plaquemines Parish, six caskets (three still in their concrete tombs) were lifted from the last bank of the Mississippi River in Pointe a la Hache, carried across the river, and found weeks later in a patch of woods on the Westbank" (p. A:1). Few instances in the history of the United States have seen such a mass grave return.[13] In Katrina's wake, approximately 1,000 bodies have been returned or positively identified, in part due to the steel name plates attached to the caskets, and in places where the storm surges were the most potent (mainly in communities along the coast) the most horrific scenes of grave disturbances occurred (Brown, 2006).

The Landscape of Recovery

The abrupt changes set in motion a series of disaster-related landscape changes that permanently altered the communities. Such disaster stricken communities can only hope that the changes do not also permanently alter the place. Place attachment becomes the foundation on which recovery and redevelopment begins for local residents. The topographical alterations serve as a testament not only to the forces of nature, but also to the human forces that shape them. As survivors struggle to rebuild and return to normalcy, the physical landscape acts as metaphor for recovery and reconstruction. The symbolic interpretation of place and human settlement becomes the primary driving force in the recovery effort. How humans respond to environmental insults and how they ultimately decide to navigate through the newly configured landscape are predicated on the symbolic interpretation of the physical landscape.

The basic premises of the symbolic approach is that *recovery begins when people act on the basis of meanings they attribute to events and conditions to develop a social response to natural or technological hazards based on their appraisal of interpretative frames of the environmental information they process, which is heavily rooted in a sense of place.* Although the environmental knowledge can come through a variety of sources, one's attachment to place forms a salient part of the self that is used to forge a post-disaster identity—

namely as a surviving rebuilder. This identity becomes important because it is connected to the land. As the water recedes, the debris is removed, a sense of normalcy is reestablished, and the social structure reemerges.

> Violently and radically, [Katrina] transformed the physical landscape, in a way exacerbated by 300 years of man's tinkering with the deltaic environment. It damaged the built environment to such an extent that vast urban expanses may have to be scraped to the ground, reconceived, and constructed anew. It thoroughly diffused and rearranged spatial distributions of every imaginable phenomenon, not within years or decades, but within hours and days. Katrina's affects will remain evident in the cityscape, both subtly and dramatically, for generations. Most significantly, the calamity will forever alter peoples' perceptions of New Orleans as a place. (Campanella, 2006b, p. 62)

Summary

Nearly two years after Hurricane Katrina and more than 200 years since it was founded, New Orleans continues to be shaped by its landscape. The need to control the forces of nature by technologically reconfiguring the topology has been met with disastrous consequences for the southern Louisiana coastline and the way of life lived by the people of this region. As the alterations to the landscape have persisted over the past two centuries, the need to explore and exploit the environment has accelerated at an unprecedented pace.

> The shifting of population and jobs from the central city to the outlying parishes resulted in a sprawling development pattern—quite remarkable for a region so constrained by natural barriers. Rather than [densely rebuilding] in New Orleans, the region [must seek other ways to sustain its population]. Density—the number of housing units per square mile—barely changed at all in the city of New Orleans over the last 30 years, increasing just 2 percent between 1970 and 2003. The end result: The New Orleans metro was consuming land at a much faster pace than its population growth appeared to warrant. (US Census, American Community Survey, 2003, as cited in The Brookings Institution, 2005, p. 10)

The purpose of a levee seems obvious—to keep the Mississippi River from spilling over its banks. But human intervention in the form of levee construction and the building of canals throughout the Mississippi River delta has not allowed the river to behave as it normally would. Lewis (2003) sums it up well: "The delta is like a bank account where there are constant withdrawals but nobody is making deposits anymore" (p. 167). Unless the river is allowed to undergo its natural replenishment process or the sediments are restored, the Mississippi delta, the lower Louisiana wetlands, and the marsh areas may no longer exist. Ironically, the levees that were built to protect property, save lives, and bring economic prosperity to a poor region of the country are potentially aiding in its destruction and the destruction of a way of life unique to the landscape that forged it.

Notes

1. In the time of Louis XIV, maps were more than mirrors of the natural world. For France during the early modern period (1666-1789), map makers sought relationships with the government, even patronage of the King, which allowed them to access more accurate information and create maps with some sense of scientific authority. This also helped interlock the relationship between government and geography and state interests (Petto, 2007). The national interests of France are expressed in its drive to chart places and firmly establish its claims to land masses in the New World. Embedded in these maps were implicit aspects of power, control, and geography or social spaces of geographic knowledge (Petto). During this period, no other cartographer's work was as important as Guillavme Delisle and his 1718 Carte de la Louisiane, which represented the most accurate data of its time and which France used to bolster its claim to the land. However, the maps of Delisle were not without controversy. The second chapter of Petto's (2007) seminal work details the time when France was considered the King of Cartography and the political nature of the maps contained information about the Louisiana colony:

> In 1718, with peace treaties signed . . . Delisle published his map of Louisiana with territorial limits that led to a volatile cartographic dispute with England and Spain. Delisle's work, Carte de la Louisiane et du cours du Mississipi, became the mother map (or source map) for all succeeding maps of the Mississippi because of its accurate depiction of the lower Mississippi river and the surrounding areas. He based his map of Louisiana (1718) on the most recent information from Soupart's survey and Le Maire's works (See Giraud, 1966). In a country whose boundaries were yet to be clarified, maps became powerful political tools in the international arena and influential in producing internal commercial propaganda. Domestically, Delisle's map joined the publicity campaign launched by John Law and his new Company of the West that took over the Louisiana monopoly from Antoine Crozat in 1717. (See Giraud, 1966)

"Internationally, Delisle's map of Louisiana generated a boundary dispute that lasted for at least fifteen years. In the map, Delisle extended the areas under French control in direct opposition to the English claims by pushing the English colonial border further east than the Appalachian frontier and then further aggravated the English by claiming that Carolina was named for the French King Charles IX" (Petto, 2007, p. 105).

2. Jean-Baptiste de Bienville (the father of Louisiana) outlives all of the earlier founders and has several distinctions in Louisiana history.

3. The actual founding of the city took place on Mardi Gras day, March 3, 1699, by the Le Moyne brothers, Iberville and Beinville. After offering thanks and worship, the explorers and their crew marked the day with a modest feast (Richard, 2003).

4. The settlers of the mid-1700s found the region to be "a place of seemingly endless, interconnected marshes, swamps, and bayous, with little soil in sight. Cat-tails, irises, mangroves, and a wide variety of grasses thrive in the delta's soggy environment" (Kelman, 2003, p. 4). The region proved to be inhospitable for the early settlers. Nonetheless, economic incentives were abound, and speculators in both France and other parts of Europe sought to get rich by way of the new colony. The river and rich marshland with a seemingly endless landscape of natural resources led to its location. In an attempt to better describe the physical landscape and lay better claim to the region, "the publication in Utrecht of Father Louis Hennepin's *A New Discovery in a Vast Country in America*, a voyage narrative of the Upper Mississippi, was first published under the title Description

of Louisiana in France in 1683. Which urged William of Orange to take possession of Louisiana" (Brasseaux, 1995, pp. 153-154).

5. A *faubourg* is an old French word that means suburb.

6. From the time French engineer Dumont de la Tour outlined the need for a levee in 1717 to protect New Orleans from Mississippi floods, an epic struggle had evolved between humanity and nature over control of the rich alluvial soil along the river. To the founders and early settlers of New Orleans, levees appeared as the most practical solution to the inundation problem. For over a century, they constructed walls of soil and sand to keep the waters of the river from intruding onto their property. It seemed they were successful. After 1815, Americans began a campaign of draining swamps, closing natural outlets, and building more levees along the length of the Mississippi River. A flood in 1828 spurred another intensive levee-building campaign. The government of New Orleans and parishes all along the river passed laws regulating the dimensions of levees, established taxes to pay for their construction, and prescribed rules for their proper care. By the 1840s, levees extended sporadically from New Orleans to the mouth of the Ohio River. Despite these efforts, a flood in 1844 broke through and ravaged plantations in Arkansas, Mississippi, and Louisiana. To rescue civilization from the river, people of the Mississippi delta looked to engineers to find a solution (Pabis, 2000).

7. Breaks in the levees occurred in every major flood. Unfortunately, engineers could only guess what effect these crevasses actually had on the Mississippi. "In the floods of 1849 and . . . 1850, the levees just north of New Orleans broke and a rush of water flowed into Lake Pontchartrain, located only a couple of miles away (Pabis, 2000, p. 68).

8. The natural risk of flooding from the lake is increased somewhat by a series of manmade canals between the Mississippi and the lake and between the lake and the Gulf (as noted above, the lake was once part of the Gulf). Some channels were dug to increase river commerce and others were dug to control flooding. For example, the Inner Harbor Navigation Canal (IHNC) connecting the Mississippi River to Lake Pontchartrain was completed in 1923. In the mid-1960s, a 500-foot wide and 36-foot deep channel from the IHNC through Lakes Pontchartrain and Borgne to the Gulf of Mexico (a distance of some 76 miles) was completed by the Army Corps of Engineers. Called the Mississippi River-Gulf Outlet (MR-GO), the intention of this canal–channel system was intended to facilitate shipping by reducing the distance to the Gulf (although the older part of IHNC is evidently too small for modern vessels). Together, the natural and manmade channels from the Lake to the Gulf allow storm surges from tropical storms and hurricanes to reach the city, and the canals into the city itself provide new avenues for those storm surges to swamp New Orleans (Congleton, 2006).

9. "Yet none of this counts for much once the lens is properly widened. It is essentially beyond challenge that Katrina would not have destroyed New Orleans had it struck the state of Louisiana as it looked on maps in 1750 or 1850 or even 1950. Thirty-six years earlier, Hurricane Camille struck the Louisiana coast with winds stronger than Katrina, but the storm surge did not reach many of the areas destroyed by Katrina. What happened in the intervening years was, of course, catastrophic land loss triggered by subsidence and canal-building. The loss of wetlands alone essentially doomed the city. Thousands and thousands of acres of marsh grass once blanketed everything south and east of New Orleans" (Tidwell, 2006, p. 25).

10. Tidwell (2006) maintains, "The calamity of Katrina, on the other hand, was probably the most widely predicted 'natural' disaster in human history. Anyone with even passing knowledge of the situation prior to August 29, 2005, knew that Katrina was coming. It was reported by journalists, described by politicians, discussed at academic conferences, simulated on computer models, outlined in government reports, and routinely predicted by every last Louisiana fisherman I ever met" (p. 29).

11. Within a month, with standing water remaining in areas and residents unofficially trickling back into the city, a new evacuation order was issued. The new storm, Hurricane Rita, posed a significant threat to life. Hurricane Rita passed well to the west of New Orleans and brought a storm surge that breached the already weakened levee system and brought nearly 8 feet of water back into neighborhoods in spite of the efforts to pump water from the city.

12. Cancer Alley is an approximate 100-mile stretch of land that is described as a pollution-ridden industrial corridor, with several major petroleum refineries found within the area. It is claimed that the area's cancer rates are higher than average compared with the rest of the country (See also Bullard, 1990a, 1990b, 2000). Bullard's (1990a) landmark work, *Dumping in Dixie*, reviewed both sitting decisions and community mobilization in southern Black communities and found strong evidence of racial disparity. Sociologist Beverly Wright and others documented the rise of the petrochemical corridor between Baton Rouge and New Orleans and its impact on poor African American "fenceline communities" (Adeola, 1998; Pastor et al., 2006; Wright, Bryant, & Bullard 1996).

13. The only other such remains return occurred in 1993 when 800 graves were disturbed during the epic flooding of the Midwest.

Figure 1.1 This picture from Slidell, Louisiana, was taken after the passage of Katrina and illustrates the power of the storm and its destructive force. Courtesy of DeMond S. Miller.

Figure 1.2 This picture taken in Slidell, Louisiana, depicts the amount of debris that the storm littered across the landscape. As seen in the picture, boats and housing materials were strewn across properties as if a landfill had been opened in the street overnight. Courtesy of DeMond S. Miller.

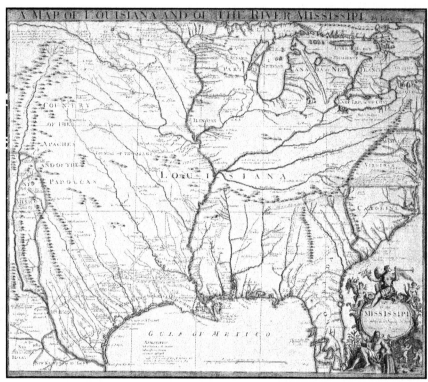

Map 2.1 This map of Louisiana was drawn in 1718 or 1719 by John Senex. As can be seen in the map, Louisiana extended in the east to the Appalachian Mountains, in the west to Spanish-controlled New Mexico, in the south to southeast Texas, and in the north to Canada. Reprinted with permission from the Historic New Orleans Collection.

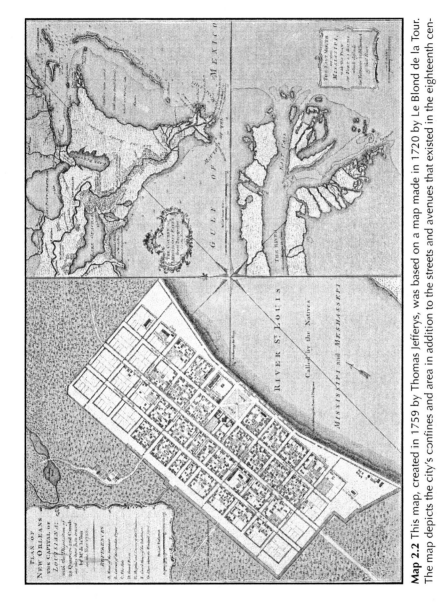

Map 2.2 This map, created in 1759 by Thomas Jefferys, was based on a map made in 1720 by Le Blond de la Tour. The map depicts the city's confines and area in addition to the streets and avenues that existed in the eighteenth century. Reprinted with permission from the Historic New Orleans Collection.

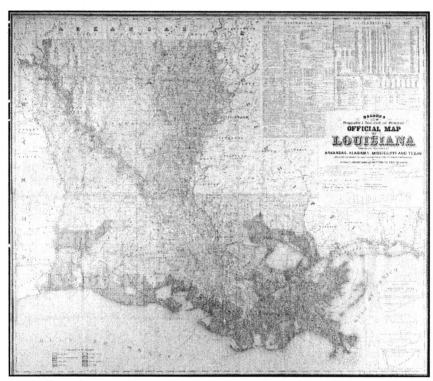

Map 2.3 This map, created in 1895 by William J. Hardee, illustrates river width and depth points, high and low water data, and information about levee heights. There is statistical and chronological information about Louisiana in the top right-hand corner, and the map is color coded (in the original work) to illustrate land use types such as alluvial, swamp, marsh, pine, upland, hills, and bluffs. Reprinted with permission from the Historic New Orleans Collection.

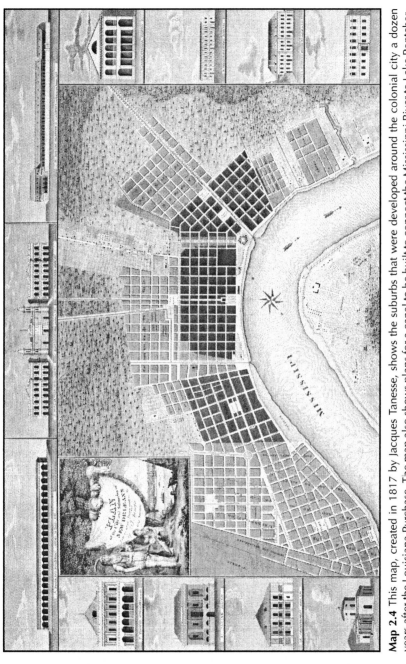

Map 2.4 This map, created in 1817 by Jacques Tanesse, shows the suburbs that were developed around the colonial city a dozen years after the Louisiana Purchase. The map also shows plans for a canal to be built to connect the Mississippi River to Lake Pontchartrain in 1807, but the canal was never built. Reprinted with permission from the Historic New Orleans Collection.

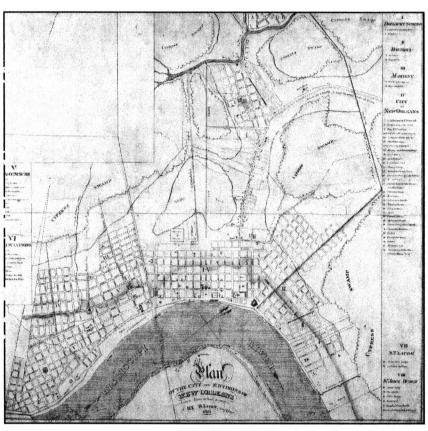

Map 2.5 This map, created in 1816 by Barthemlemy Lafon, depicts the city's eight subdivisions: Annunciation, City of New Orleans, Daunois, Declouet Suburb, Marigny, St. Claude, St. John Burgh, and St. Mary suburbs. The map also shows plans for a canal to be built connecting the Mississippi River to the Carondelet Canal via what is now Canal Street, but the canal was never built. Reprinted with permission from the Historic New Orleans Collection.

Map 2.6 This map, created in 1878 by Thomas S. Hardee, depicts the city's late nineteenth-century development. It also illustrates the level of canal building that had taken place up to this point, as seen in the canals reaching from Lake Pontchartrain to the Mississippi River. Reprinted with permission from the Historic New Orleans Collection.

Chapter 3
The Cultural and Economic Landscapes

> Nature had done what "modern life," with its relentless pursuit of efficiency, couldn't do. It has done what racism couldn't do and what segregation couldn't do either. Nature has laid the city waste—with a scope that brings to mind the end of Pompeii.
>
> —Anne Rice (2005, p. 14)

The unprecedented topographic changes of the Gulf Coast region, the massive demographic shifts brought on by evacuations, and the distress felt by the citizens have extensively changed the cultural and economic landscapes of the region. The cultural and economic landscapes that rest within a specific geographic location are dependent on each other for their development. The economic landscape controls many features of the social class structure, the interactional potential placed on individuals, and the tangible outputs of the society, thereby influencing the development of the cultural landscape. However, the cultural landscape also influences the economic landscape in that culture dictates what commodities are acceptable for manufacture and distribution in an area, directly affecting the economics associated with a place. The cultural and economic processes that have unfolded over the past two centuries shape the physical landscape and the lives of the people of New Orleans. Moreover, the city's mixture of individual cultures come together in a space that its residents endow with special meaning, which in turn forms a unique place that people from around the world seek to experience.

However, in the aftermath of Katrina, most of the population that contributed to the city's world-renowned culture remains in diaspora over two years after the disaster, with former residents in nearly every state. In rebuilding New Orleans, the losses beyond relief and reconstruction dollars (i.e., cultural contributions) must be considered, which is something few Americans are able to do (Marsalis, 2005). Although the standard paradigm for natural disasters is that an initial brief period of stymied economic activity ultimately gives way to reconstruction efforts that will fully compensate for the initial economic loss, recovery from Katrina is not likely to follow this basic script of economic and cultural development (Dekaser, 2005). The city's development will most likely be hampered by the combination of the sheer magnitude of the disaster, the failure of government response at all levels, the weeks of stagnant flood waters, and the delay in returning the port to its pre-Katrina state.

Historical Effect of Race and Culture in New Orleans

The exchanges of possession the Louisiana territory went through between France and Spain[1] has made the people who have lived there subject to a culture that took shape amid the physical landscape of the city and the Old World perceptions of Blacks and indigenous tribes. Moreover, New Orleans' key position as a port city with access to the American Midwest contributed to the multiethnic exchange of ideas, language, and culture attributed to many other trade routes in history.

With the combination of the climate and topographic features, such as the bayous and marshes surrounding the city, the landscape allowed the initial fledgling colony to use its location as both a strategic trading post and a viable agricultural supplement to the French Empire. It was for these two reasons the port city appeared attractive to both international empires and immigrants seeking opportunities and a new life. Although the colony had obvious strategic appeal, it had initial difficulty attracting a variety of immigrants. The population resulted in stationed soldiers, indigenous peoples, settlers seeking fortune, and slaves. This contributed to the mixing of cultures even at this early point in the colony's history. By the time the Spanish colony was ceded back to France, interracial mixing had become so common that the French had to adopt a social classification system[2] similar to the racial hierarchy used in their St. Domingue colony, which is currently known as Haiti (Dominguez, 1986).

The mixing of different races directly affected the cultural development of the city and the social interactions that have taken place. The Spanish census of New Orleans conducted in 1778 counted 3,059 people and comprised 51 percent who were white, 31 percent who were slaves of pure African blood, 8 percent who were free people of mixed blood, 7 percent who were slaves of mixed blood, and 3 percent who were free Africans of pure blood (Robichaux, 1973, as cited in Campanella, 2006a). Due to Western perceptions of ethnocentrism, racial segregation played a major role in the cultural development of New Orleans. During the late 18th century, the Spanish authorities made racial mixing illegal (Alcubilla, 1891, as cited in Dominguez, 1986), which was already a moot point due to the amount of racial mixing that had taken place prior to the establishment of the law (Miller & Rivera, 2007b).

The enactment of the Spanish law was more actively followed by the White elites of New Orleans, who were concerned with issues of inheritance and family titles. The law also affected Black segments of the city's population. According to Kephart (1948), the enactment of the Spanish law caused segments of the Black population who shared physical characteristics with Whites (gained through prior racial mixing) to socially disassociate themselves from more "pure blooded" Blacks. Although the Spanish law tended to be followed more prevalently by elites in both racial communities, Kaslow (1981) explained that racial mixing continued to exist in the lower echelons of New Orleans society.

Tregle (1992) explains that it was not until the Americans gained authority over the territory that segregationist policies and practices were officially followed. Until the transfer of the Louisiana Colony to the United States, Blacks and Whites resided alongside one another in life and death[3] in New Orleans, and both had access to economic prosperity.[4] However, when the Americans gained authority, economic and social access was eventually denied to all Blacks. Tregle (1992) further asserts that it was important not only to be a freedman during American control but also to be devoid of any African descent. American control of the city and the rest of the territory eventually motivated Blacks sharing enough physical characteristics with Whites[5] to further disassociate from "more pure" Black communities, breaking from "Black culture" and accepting "White culture" (Tregle, 1992).

Somers (1974) explained that Blacks and Whites were capable of living in close proximity for a short time after Americans gained control and prior to the enactment of Jim Crow Laws[6] because local Whites did not completely adhere to a mentality of Black subjugation. Even though American administration brought with it a more racialized social landscape, between 1837 and 1860, the city attracted more immigrants than any other American city (with the exception of New York), totaling more immigrants than Boston, Philadelphia, and Baltimore combined (Treasury Department, 1889, as cited in Campanella, 2006a). By 1850, the city contained more ethnic groups of significant size than any of the other twenty-nine major American cities of the time. At the time of the 1850 census, New Orleans contained seven separate groups that comprised at least 5 percent of the total city's population when no other city in the county had more than five ethnic groups of significant size (De Bow, 1854, as cited in Campanella, 2006a).

During the second half of the 19th century, the city experienced an influx of Black migration in addition to an increase of European immigration into the area. Migrants from Ireland, Germany, and Italy came by the thousands, bringing parts of their cultures with them. European migrants built churches, convents, schools, and orphanages and expanded the city in all directions, creating new neighborhoods (Rice, 2005). Although there was always a European presence within the city, as evidenced by the Old World architecture indicative to the city, the increased European presence aided in further diversifying the cultural landscape (Kaslow, 1981). Although there was an increase in the ethnic White population that would have expectedly increased racial tensions (Kaslow, 1981), Blacks were able to enjoy a higher social standing in urban society than other similar urban localities in the South due to an escalating Black population and the practice of "White flight" (Somers, 1974). This illustrated their ability to be successful under strenuous circumstances; however, it changed with the enactment of racially discriminating laws in the late 19th century that disenfranchised millions of Blacks living in the South. During the late 19th and early 20th centuries, generations of Blacks were born into a Jim Crow society that perpetuated inequality.

Landscapes of Contrast: Socio-Economics at Play

In New Orleans, the socio-economic contrast between the rich and poor and the Blacks and Whites has had profound effects on the cultural development of the city. This contrast, primarily caused by segregation and other race-based policies, has both directly and indirectly regulated people's social mobility, life experiences, and in some cases, the potential for life expectancies. As discussed earlier, this contrast has affected the development of place attachment. The physical landscape's influence on the social and economic conditions of the city remains a driving force in the development of social and economic conditions both before and after Katrina. The major question exists: With such an abrupt change in the physical landscape, will there be a reestablishment of traditional culture in the metropolitan area, or will such drastic changes force the development of a cultural landscape with an entirely different trajectory?

With an expanding Black population in the city throughout most of the 20th century and the existence of racist public policies that reinforced residential and social segregation, some New Orleans citizens remained adversely affected due to a cultural and economic legacy tainted by racial segregation long before Hurricane Katrina. According to Dreier (2006), the city was not only one of the poorest cities in the United States but also one of the most ghettoized. This was a result of the Great Mississippi Flood of 1927; however, in prior periods, the city was a prosperous national port and a developing city (Kelman, 2003).[7] Pettit and Kingsley (2003, as cited in Dreier, 2006) explained that in 2000, 23 percent of the poor people who lived in the city were in "high-poverty" neighborhoods, and approximately 40 percent of the population lived below the poverty line. In recent years, the city's residential segregation and poverty concentration has only increased. According to Berube and Katz (2005), the average Black resident in New Orleans lived in a neighborhood where 82 percent of the population was Black. Residential segregation was so prevalent in New Orleans prior to Katrina that estimates contend that for there to be an equal distribution of Blacks and Whites in every neighborhood, 69 percent of all Blacks in the city would need to move (*Sortable List of Dissimilarity Scores,* 2005, as cited in Dreier, 205). Gieryn (2000) maintained that social classifications are built into everyday material places and landscapes, such as by concentrating specific economic and racial classes into defined geographic areas. The social classifications confront imposing and constraining forces:

> Place sustains difference and hierarchy both by routinizing daily roads in ways that exclude and segregate categories of people and by embodying in visible and tangible ways the cultural meanings variously ascribed to them. (Gieryn, 2000, p. 474)

The segregation of Blacks into generally low-income neighborhoods has directly affected the group's interactional potential.[8] The segregation of Blacks within New Orleans has "routinized" the phenomenon so much so that prior to Katrina, it would have been culturally expected within the city for Blacks to live in low-income neighborhoods. This expectation was not uncommon throughout the city and seen as a fact of life and a cultural characteristic of the indicative landscape.

According to Popkin, Turner, and Burt (2006), city planners have systematically excluded the city's "better-off neighborhoods" from containing assisted housing, further contributing to the concentration of poverty in communities already containing a large percentage of residents living below the poverty line.

The city's concentration of poverty persisted and eventually worsened. Between 1990 and 2000, through urban planning, the city geographically concentrated its poverty-stricken residents from living in fifty-nine census tracts to forty-seven (The Brookings Institute, 2005). Additionally, concentrations of poverty coincided with residential neighborhood segregation practices; Treme/Lafitte, Central City, and Gert Town all have approximately half of their population living below the poverty line, whereas Old Aurora and the Lakeview neighborhoods had poverty rates below 10 percent (The Brookings Institute).

These areas of concentrated poverty have high levels of home ownership in contrast with a relatively low homeownership level for residents not living in poor neighborhoods (Zedlewski, 2006a). In the Lower 9th Ward specifically, nearly two-thirds of the residents were homeowners, and many had lived in the same family home for generations (Popkin et al., 2006, p. 19). Additionally, Popkin (2006) explained that only 52 percent of residents living in the Lower 9th Ward had a mortgage or homeowner's insurance that lenders required at the time of Katrina. The presence of such a high level of homeownership in an otherwise poor area points to the strong degree of place attachment felt by families who had lived in the area or in the same location generation after generation.

In tandem with residential segregation was the presence of low levels of employment and low-income earning jobs among Blacks (Cutler & Glaeser, 1997; Holzer & Lerman, 2006). In 2004, the city experienced an unemployment rate of 12 percent, which was twice the national rate (Holzer & Lerman, 2006). According to Holzer and Lerman (2006), approximately 13 percent of the city's work force was employed in the low-wage food and accommodations industries. Moreover, service jobs accounted for 26 percent of all jobs in the city, paying on average only $8.30 per hour. Prior to Katrina, there was a relatively high percentage of self-employed adults (10 percent) working as either freelance musicians or artists (Zedlewski, 2006a), which accounted for a large portion of workers employed in the city's tourism industry. Self-employment seemed to have been a preferable life choice when compared with many of the lower-paying options in and around the city.

Education levels also contributed to the city's high unemployment rate (Holzer & Lerman, 2006). The Brookings Institution (2005) reported that the New Orleans metropolitan area ranked 80th out of the 100 largest metropolitan areas for college attainment, which was 23 percent in New Orleans. Also according to Zedlewski (2006a), 28 percent of teens between the ages of 16 and 19 lacked a high school education and were unemployed. Glaeser (2005, as cited in Holzer & Lerman, 2006) contended that due the labor force's limited educational attainment and cognitive skills, most poor and disenfranchised residents were forced into low-wage industries. The lack of a minimum wage in Louisiana, an increasing unemployment rate, and a declining population aided in the social degradation of some of New Orleans' poorer neighborhoods prior to the storm (Holzer & Lerman, 2006). Such wide-spread disparities among Whites and Blacks were indicative to Black social conditions prior to the storm. The contrast of Black neighborhoods with adverse social conditions and neighborhoods of the more affluent White population has resulted in the instillment of *separate lifeworlds* in the city (Dawson, 2006).

The idea that Blacks and Whites have developed separate lifeworlds within New Orleans is a testament to the extent to which segregation has played a role in the experiences of people from both races.

> Blacks and Whites are segregated into largely separate lifeworlds. By this we mean that if the lifeworld is seen as a basis for consensus, communicative action, shared comprehension, and therefore, social integration and reproduction, then these are not shared between Blacks and Whites. Empirically, the consensus, shared comprehension, mutual understanding, and social integration that Habermas details as necessary in order for different groups to share lifeworlds is lacking in the racially hypersegregated society that is the United States. (Dawson, 2006, p. 241)[9]

Development of these separate lifeworlds results in the production of a fragmented civil society and racially separated public places and social spheres (Dawson, 2006; Habermas 1992[1996]), which is manifested in the formation of different world views (Dawson, 2006). By developing different world views, both races are able to experience the same phenomena, but their interpretation of the event will be different; therefore, each will assign different significance to the event. This is important because the significance of events in specific geographic locations influences an individual's place attachment and sense of place. World views also have the ability to perpetuate long-standing social conditions, which is indicative to the perpetuation of segregation by Whites in New Orleans because, in their world view, they are content with social conditions as they are.

The development of one's lifeworld is more dependent on life chances and life choices. Life chances are "the chances throughout one's life cycle to live and to experience the good things in life" (Miller, 1992, p. 317). Miller asserts that "the family to which one is born is the most significant variable in predicting an individual's life chances. Gender and race are also established factors in affecting life chances in American society" (Miller, 1992, p. 317). Life choices are the decisions people make in accordance with their life chances that will either increase or decrease their life chances. If one were to look at life choices and life chances, he or she could see the relevance of race. Life choices are bound by a person's life chances. "The disadvantaged African American New Orleanians who remained behind for reasons other than an inability to secure some means of departure stayed because they could not bring themselves to abandon their meager possessions, which were contained in their modest domiciles and which stood as all that they possessed in their world beyond their physical being" (Young, 2006, p. 204).

Other segments of the city's poor, such as the elderly, the disabled, or single mothers, contained poverty rates that were all higher than both the national and Louisiana's averages. According to Zedlewski (2006b), approximately one in five elderly people living in the city were living below the poverty line, which was twice the national average. The social condition of the elderly was made worse by the fact that 56 percent of the city's elderly population (those aged 65 years and older) reported some type of disability in 2004 (Zedlewski, 2006b).

These statistics seem to support the notion that poverty and poor health among elderly individuals is connected not only in New Orleans but everywhere (Smith & Kington, 1997; Zedlewski, 2006b).

The city's number of working-age individuals with disabilities was also relatively high in comparison with the national averages prior to the storm. Additionally, like other population segments, poverty ranked high among the working-aged population. Contributing to the poverty rate of this group was the fact that approximately one-third of these individuals lacked a high school education, which limited their access to employment (Zedlewski, 2006b). Moreover, in otherwise better social conditions, working-age adults with disabilities are less likely to be employed, so it is not surprising that this population's employment rate was 32 percent, falling below the national average (Cornell University, 2004; Zedlewski, 2006b).

The number of single mothers, another major segment of the city's poverty stricken population, supports social theories correlating single parenthood and poverty (Corcoran & Chaudry, 1997; Zedlewski, 2006b). According to Maximus (2002, as cited in Zedlewski, 2006b, p. 65), only one in five low-income families who resided in Orleans Parish prior to the passage of Katrina included both parents. The Bureau of the Census (2004) indicated that 54 percent of single parent families living in New Orleans were poor, which was far above the national average of 38 percent. Further adding to this population segment's social condition was the fact that 79 percent of never-married, single parents were African American (Maximus, 2002), and approximately half of these parents had not completed high school (Zedlewski, 2006b). The lack of education and the personal demands placed on single parents, in combination with other social conditions, contributed to high levels of unemployment and poverty indicative to New Orleans prior to the storm.

> Without a high school diploma and job skills, and sometimes in the grip of poor mental health and substance abuse, a large share of the single parent[s] in New Orleans were hard-to-serve parents with serious barriers to employment. Many were jobless. . . . For these parents, finding childcare present[ed] a considerable challenge. (Knox et al., 2003, as cited in Zedlewski, 2006b, p. 65)

Among this group, barriers to employment significantly contributed to the city's high poverty rates, making subsistence all the more difficult.

Social Services and Economic Limitations

With the profuse existence of social strife that was inherent prior to Katrina, one is forced to wonder about the presence and abundance of social support programs. In many circumstances, federal, state, and local governments supplement resources to poverty stricken individuals and families in an effort to act as a safety net; however, Louisiana has historically spent relatively little state resources in this area (Zedlewski, 2006b). Although Yilmax, Hoo, and Nagowski (forthcoming) argued that Louisiana's lack of fiscal capacity has limited, and still limits, its ability to invest in these financial safety-net ventures, the state's inadequacies do not change the fact that residents direly need federal financial resources. This does not mean that the state was totally lacking in supportive financial services to its population. Prior to the storm, 3 percent of families in New Orleans received aid from the state's Family Independence Temporary Assistance Program, which provided two hundred dollars per month for a family of three; moreover, this aid, which was distributed to poverty stricken families, was almost completely (99 percent) distributed to families headed by single mothers (Zedlewski, 2006b).

Two other aspects of social services that are usually provided by the state or federal government are food and housing assistance. In reference to food assistance programs, approximately 10 percent of families in the city received food stamps, which reflects the area's high poverty levels (Zedlewski, 2006b). The high rate of food stamp use within the city was not surprising to observe since Louisiana had the sixth highest Food Stamp Program participation rate nationally in 2003 (Kastner & Schirm, 2005; Zedlewski, 2006b). Moreover, families received assistance from the Women, Infants, and Children program, which was responsible for providing 83 percent of school-age children with free or reduced-price school breakfasts and lunches in 2002 (Maximus, 2002; Zedlewski, 2006b). However, even with these food programs in place, Maximus reported that in 2002, one in five low-income families went hungry, and another quarter were food insecure without experiencing hunger. Housing assistance prior to the storm also seemed limited, with nearly half of low-income families with children paying rent without assistance (Zedlewski, 2006b). Furthermore, Zedlewski (2006b) notes that more assistance was needed prior to the storm because escalating housing and rent costs equaled or exceeded 40 percent of a household's income for over one-third of the city's low-income families in 2002.

As discussed previously, prior to the storm, a large portion of the city's residents had several personal barriers to entering the workforce, and in some circumstances, entire segments of the population were institutionally marginalized, further complicating their ability to get jobs. However difficult it was for these segments of the population to obtain employment, the situation was made worse because the overall job market of the city had been on the decline for a period of time prior to the storm. This had detrimental effects on employment for even those who were qualified for jobs. According to The Brookings Institute (2005), between 1970 and 2000 the metropolitan area experienced an economic structural change that resulted in the net loss of 13,500 manufacturing jobs, a job decrease of 23 percent. During this time, the manufacturing and transportation industries accounted for 12 percent and 10 percent of employment in the area, respectively, but by the year 2000, both industries only accounted for 6 percent of employment (The Brookings Institute, 2005, p. 11). As manufacturing and transportation industries declined, there was immense growth in the service sector, which grew by 136 percent between 1970 and 2000 (The Brookings Institute, 2005, p. 11).

The evolution of the region's economic landscape also had a drastic effect on the metropolitan area's residents. With an increase in the service sector, fewer well-paying jobs were available to individuals without a college education. The Brookings Institute (2005, p. 11) reported that in 2000, the average annual pay for manufacturing jobs was 62 percent higher than the pay for individuals working in the service sector; however, access to this sector was limited to the majority of the area's residents due to a lack of educational qualifications. Excluding government jobs, the largest non-farm sectors of the area's economy were retail, accommodation and food service, health care, professional and technical services, and other services such as repair and maintenance, personal services, and laundry. Of these sectors, four paid below the national annual average of $43,061 for non-farm sectors, and accommodation jobs averaged no more than $19,131 annually (The Brookings Institute, 2005, p. 11).

Many of the region's more prosperous industries, considered export industries, were also declining between 2000 and 2004. In 1970, the region was known for its tourism, oil and gas extraction, oil refining, chemical manufacturing, the port and related transportation industries, waste treatment, ship and boat building, commercial banking, higher education, corporate headquarters, and insurance industries (The Brookings Institute, 2005, 2005, pp. 11-12). However successful they were in the past, between 2000 and 2004, several of these core sectors of the economy shrank. In addition to the oil and gas extraction, chemical manufacturing, and the port and related transportation industries, several high-paying industries in the area lost a significant number of jobs after the year 2000. "The New Orleans region's shift toward lower-wage service sectors and recent inability to produce jobs in high-value industries have each limited the quality of the opportunities available to New Orleans-area workers, both well-educated and less educated" (The Brookings Institute, p. 12).

Renowned Cultural Characteristics

Despite the debilitating socio-economic conditions present in New Orleans, a culture that is world-renowned and attractive to all social classes has developed. The culture and arts known to be indicative of New Orleans are valuable expressions of the people who live there (Jackson, 2006). Although the people of New Orleans experienced social hardship and racial discrimination before the storm, the ability of the people to produce a culture unique and rich in the arts is a testament to those who lived there. As expressed previously, a majority of the people who lived in the city prior to the disaster were either officially or unofficially employed in the tourist industry but were more specifically employed as freelance musicians and artists. What is interesting about the city and its people is the fact that, despite the social situations that have evolved over the centuries, the people seem to have a more inclusive concept of arts and culture than in other similar places (Jackson, 2006).

> Vibrant arts and cultures in New Orleans shined the one bright beacon on an otherwise depressed landscape for low-income families. Indeed, many of the cultural practices and traditions that made New Orleans famous can be traced back to the city's poorest citizens and their ancestors. Arts and culture were key to New Orleans' unique character. (Zedlewski, 2006a, p. 7)

Moreover, the residents of the city share something that is lacking in other places across the nation—recognition that "root culture" matters (Jackson,

2006). Mardi Gras is one of the primary illustrations of "root culture" recognition. By taking part in and observing Mardi Gras festivities, locals and outsiders are sharing in the passage of cultural history from one generation to the next and sharing their culture with the rest of the world.

It is the recognition of people who have lived in the same geographic location in the past and the location's transference of built places into the present that have enabled the city's aesthetics to become part of its indicative culture. On its own, the city was an aesthetic representation of the region's cultural and political past, with its combination of Spanish, French, and Victorian influenced architecture. Moreover, the city architecturally represents the entire United States' colonial history. But aside from the city's past political roots that have profoundly influenced the formation of the city's culture, the city's inherent historical ties to slavery, discrimination, social inequity, and a misrepresented underclass of individuals has contributed to some of the most acknowledged music styles and types of cuisine the nation and the world has seen.

Katrina's Effect on the Culture Landscape

The hurricane's effect on the cultural landscape of New Orleans cannot be underestimated. Katrina forced thousands of people from their homes and dispersed them across the country. However, in the French Quarter of the city, bars and restaurants on Bourbon Street and many of the city's museums and galleries were open soon after the passage of the storm (Robinson, 2005). The French Quarter gives the city hope that its culture can return (despite the widespread destruction) due to its ability to return to business as usual, which is what defined the city. According to Hutter (2007), the French Quarter and Bourbon Street were what outsiders viewed to be New Orleans' culture.

> home of the Mardi Gras, of fine food and drink, of music—jazz and zydeco—of Brennans for breakfast—of Café Du Monde and late night/early morning beignets and chicory coffee—of po' boys and crawfish—of the stately mansions of the Garden District—of Audubon Park—of Jackson Square and buggy rides and steamboat river cruises—of the aquarium and of professional football at the Superdome. (p. 434)

New Orleans' culture, as viewed from the perspective of an outsider, is about enjoying life through the senses, which places significant emphasis on the city's indicative styles of music, food, drink, and promiscuity (Falk, Hunt, & Hunt, 2006). According to Falk et al. (2006), it was the city's violation of norms indicative to most other locations that drew people to New Orleans, which was "a city mixing historic charm, public spectacle, and theme park qualities" (p. 119). In the more affluent historical areas, culture and daily life proceeded as normally as could be expected after the storm, but for other more poverty stricken areas, Katrina was not as kind. In the Lower 9th Ward and other neighborhoods close to the French Quarter, many residents lost their possessions, family members, friends, and homes (Jackson, 2006). Adding to their emotional loss is the fact that many of them were temporarily relocated, but the question still remains as to whether or not many of these people will return to New Orleans at all (Holzer & Lerman, 2006).[10]

The implications of the more affluent areas returning to normal and being untouched by the storm have not been fully realized; however, there is some indication that the results will be far reaching. There have been discussions of plans to redevelop the area for the more affluent segments of American society, thereby permanently displacing the less affluent. Also, some share the notion that rebuilding the city as a rich tourist destination will attact visitors to the city because there will be a lack of "less desirable" elements of the population and segments of the urban landscape (i.e., fewer poor individuals results in fewer poverty stricken neighborhoods and social conditions associated with poverty).

What has escaped the authors of these ideas is the fact that most of the people who contributed to the city's cultural and economic vitality and its development were those who did not live in these more affluent areas. By taking a gentrification approach to rebuilding the city, many long-time residents would be forced out and their cultural contributions would leave with them decimating the cultural landscape. The reasoning behind this logic is that these more affluent neighborhoods have been able to recover from and offer the notion that the dispersed portions of the population were not needed for economic and cultural vitality in the first place. However, this is far from fact.

Among the scattered people were many of the city's Mardi Gras Indian parading crews and Mardi Gras Tribes,[11] brass band members, independent musicians, and cooks. Church congregations and parishes as well as social aid and pleasure clubs, which were established to support leisure and community needs, were also dispersed by the storm (Jackson, 2006). In some situations, large portions of neighborhoods have been dispersed,[12] such as the neighborhood of Treme, which was known for its contribution to the city's music and culture. The neighborhoods have been emptied and so have the community institutions that aided in socially organizing the city (Jackson, 2006). In communities like Treme, where there are a disproportionate number of poverty stricken residents, the reconstruction of social institutions such as churches is highly questionable, especially when they must rely on their constituents for aid in the rebuilding process (Webster, 2005).

In other sections of the city that were known for their artists, art organizations' small clubs, and music venues (such as in Bywater Marigny), the ability of these people and organizations to return is limited by their ability of financially supporting themselves (Jackson, 2006). In summary, "The culture of New Orleans is very much at risk when the people who make and preserve it are scattered and living in a sea of uncertainty and when the places where artists and traditional bearers live, where they make and practice their arts forms, are largely destroyed" (Jackson, 2006, p. 57).

Katrina's Impact on the Economic Landscape

Obviously, with the displacement of so many people and the destruction of a large portion of the urban environment, which has not been seen in an American city since the destruction of Atlanta during the Civil War (Lang, 2006), there have been immense ramifications on the city's economic stability. As discussed previously, the city was losing jobs increasingly to other areas of the country in addition to the increased strain on its service sectors prior to Katrina. Even as a port city with much of the nation's agricultural shipping passing through New Orleans, reopening the port has not been a great boost to the city's economy because the heavy use of automation and containerization produces little need for workers (Lang, 2006). Moreover, Katrina may have brought the city closer to something that has been feared for years—that the Mississippi will change its course to find a new outlet into the Gulf of Mexico by possibly passing through the Atchafalaya River, which would cause New Orleans to lose its port and any associated revenues (Lang, 2006; McPhee, 1989).

Katrina caused many people to lose their jobs[13] because much of the social and physical infrastructure on which the labor market was based has been destroyed (Jackson, 2006). Had the labor pool stayed stable throughout the disaster, this would have been a serious issue of concern because it would have further limited the available jobs. However, due to the already declining job market and the displacement of so many of the city's residents to other locations, the city actually has a surplus of jobs but lacks a sufficient labor pool to support its economy (Jackson, 2006). Adding to employment issues was the occurrence of a disaster economy, or *disaster capitalism*, almost immediately after the storm, resulting in both private and public organizations moving into the area.[14] According to Klein (2005, as cited in Dyson, 2006),

> disaster capitalism is a predatory force that "uses the desperation and fear created by catastrophe to engage in radical social and economic engineering," driven by the reconstruction industry, which "works" so quickly and efficiently that the privatizations and land grabs are usually locked in before the local population knows what hit them. (p. 131)

The presence of these organizations and their employees in the area solely to take advantage of the low labor pool in an attempt to secure lucrative contracts and employ outsiders diminishes the ability of local residents who remain after the storm from gaining employment. Through disaster capitalism, the economic landscape has limited the earning potential of people living within the city, especially in impoverished neighborhoods. Although we do not argue that there is a need for outside experts to assist in the cleanup and redevelopment efforts, local residents (who in many cases have no jobs to which to return) are oftentimes left out of the loop because lucrative governmental contracts are awarded to businesses that originate and employ others from outside the disaster impact area. Klein (2006) postulated a grim future where disaster capitalism makes money out of misery by "the privatization of aid after Katrina [that] offers [us] a glimpse of a terrifying future in which only the wealthy are saved" if we as a society reinforce politicians and public policy, which allows governments to shift their responsibility to the private sector.

These companies were reconstruction oriented and enticed into the area not only by the void of other companies but also by President George W. Bush's decision to suspend the Davis-Bacon Act of 1931 and allow companies moving into the New Orleans area to pay workers less than the prevailing local wages (Dyson, 2006). Dyson explained that the President's decision had three effects: (1) it contributed substantially to the corporation's profits; (2) it further financially destabilized local wage workers; and (3) it reinforced the consequences of an already racially segmented work force.

Although the Davis-Bacon Act was reinstated after pressure from both Democrats and Republicans, the financial effect on local poor Black workers had already been felt and may have even forced vulnerable evacuees into permanent exile due to a lack of financially desirable employment (Dyson, 2006). What is needed to aid in the redevelopment of the economic landscape is more than simply jobs; what is needed is an increased labor pool composed of local residents. However, attracting former residents to return to live in the city affects the reestablishment of the economy.

According to Zandi, Cochrane, Ksiazkiewicz, and Sweet (2006), disaster-stricken regional economies are usually the beneficiaries of substantial economic aid. In the case of Katrina, private insurers are expected to pay out about $40 billion in household and business claims in addition to an anticipated $90 billion from the federal government. However, regardless of the amount of aid given to the region, for New Orleans to bounce back economically, it must overcome the obstacles of a reduction in demand for goods and services in addition to its lack of liquid labor (Zandi et al., 2006). Roig-Franzia and Connolly (2005) explained that although there has not been a shortage of employment since the storm, few prior resident workers have decided to return. This may be due to the type of work that is available, which is mostly low-wage; however, it is also likely due to the lack of available housing (Holzer & Lerman, 2006). Falk et al. (2006) explained that those people with lower incomes who have been displaced face the decision of whether or not to return to a devastated home, neighborhood, and community with poor earning opportunities and access only to inadequate services (medical, educational, and retail).

The resettlement of the city is further complicated by the inability of the New Orleans School System to adequately serve the nearly 60,000 displaced students. The issue of schools reopening is further complicated by the discovery of harmful toxins in school buildings and on playgrounds left by the flooding (Pastor et al., 2006; Ritea, 2006). The inability to recruit staff has also prolonged closures of both primary and secondary schools.

Falk et al. (2006) predicted that many people will remain in the locations that they have been dispersed to because, as time passes, their new locations will increasingly become more "home-like" to them, causing dramatic demographic changes to a delicate social structure that developed over centuries.

> These changes can be said to represent, in a condensed temporal frame, many of the same processes recognized in other contexts as elements in the global restructuring of cities. In New Orleans, as in other globalized cities, one discerns dramatic "reconfigurations of wealth, social space, and urban citizenship." (Zhang, 2002, p. 312)

> As in other displacements, people unable to return to New Orleans are losing more than just shelter. The social networks they developed over generations are also threatened. (Breunlin & Regis, 2006, p. 745)

Although this may not substantially affect the culture of these displaced persons individually (Falk et al., 2006), it will affect the cultural landscape in New Orleans.

New Orleans' Changing Cultural Landscape

The predominant presence of low-wage jobs has enticed an influx of immigrant populations expecting to receive jobs due to the already limited labor pool. This has made prior residents less inclined to move back into the area and has prompted anti-immigrant sentiments as far up the social ladder as Mayor Ray Nagin. When confronted with the notion of an immense increase in the city's immigrant populations, Mayor Nagin rhetorically remarked, "How do I ensure that New Orleans is not overrun by Mexicans?" (Radelat, 2005, as cited in Muniz, 2006, p. 15).

The increased presence of immigrants (Latinos and Asians) points to economic alterations in the cultural landscape as well as demographic changes.[15] According to the Honduran Consulate, 140,000 Hondurans and their descedents lived around New Orleans alone prior to the storm, representing the largest Latino subgroup in the area and the largest Honduran population in the world with the exception of Honduras itself (Muniz, 2006).[16] According to Muniz, these immigrants were attracted to the area because of job opportunities in construction and the hospitality industries prior to Katrina. The increased demands placed on the construction and reconstruction industries caused a deficiency in the number of available skilled workers. This fact led to the assumption that more immigrant migration into the area, especially by Latinos, will occur because officials are not asking for working or legal residence papers (Quinones, 2006). Demographic changes (i.e., a large influx of immigrant workers) also will impact the cultural landscape; as different people bring cultural and ethnic values to the city, the cultural landscape will change to be more inclusive.[17] Mayor Nagin stated that New Orleans should remain a "chocolate city," more specifically an "African American city . . . the way God wants it to be" (Dao, 2006, as cited by Hutter, 2007, p. 441). In an effort to increase worker's income and make the city more attractive to job-seeking individuals, money was subsidized to pay workers.

> The socio-economic landscape is in such disrepair, due to the dismal state it was in prior to the storm and then from the damages inflicted on its supporting infrastructure, that there has been discussion of outlaying $10 billion over the next year [FY 07] for employment, which would supply 169,000 jobs paying $40,000 salaries and $20,000 in other costs for a full year in rebuilding the city. Subsidizing employment in this way would hopefully stimulate the economy into reestablishing itself and additionally indirectly stimulating the creation of local jobs in other occupations and industries. (Holzer & Lerman, 2006, p. 12-13).

However, subsidizing jobs is a short-term solution. For the plan to have long-term effectiveness, the economy, population growth, and employment opportunities must be sustainable; city officials and planners believe that a viable economic infrastructure can be achieved through the expansion of the service sector.

Capitalizing on New Orleans' Cultural Landscape

As discussed earlier, tourism became one of the more prominent (if not the most prominent) industries in New Orleans prior to the storm. The storm has virtually created a blank slate for the city, almost as if the city was wiped from the Earth and another city was built on its ruins. Although the city has not been completely erased from the Earth, a large enough portion of the city was destroyed that people question whether the city should be rebuilt in a different location to avoid future property destruction. The storm has wiped away most remnants of the prior socioeconomic constraints of segregated housing and poverty and has left the possibility for a fresh start in the pursuit of a more equitable socioeconomic landscape; however, if housing patterns and economics change, so too will the culture of the city.

What has been proposed as a method of both creating more equitable social conditions for residents and stimulating New Orleans' economy is to expand the tourist industry. Many cities have used tourism as a strategy for urban redevelopment and revitalization, forcing local governments and the tourist industry to forge close institutional and fiscal ties to "sell" the city. By sustaining large quantities of jobs and occupations, including advertising campaigns, recognizable attractions, and diverse forms of financial investment, the tourism industry is sustained (Gothman, 2002). Each attempts to market its place through the use of imagery and themes, which usually positively affects the economy of cities (Gothman, 2002).[18]

Although true that by increasing the tourism industry there will undoubtedly be an increase in the number of jobs, it will not alter the social conditions of the city that were present prior to the storm. It is not argued that there will be a possible increase in employment in the short-term in addition to a preservation of the city's culture and sense of place by adhering to cultural revitalization plans, such as Mayor Nagin's *Rebuilding New Orleans* plan. The larger question is to what extent culture will be developed and sold to the public as opposed to being developed for development's sake.

Mayor Nagin's plan for rebuilding the city contains a cultural redevelopment component that seeks to reestablish the city's musical, visual, culinary, architectural, literary, and graphic arts due to their ability to draw tourists to the area.[19] This plan calls for approximately $648 million of investment over a three-year period to jumpstart the redevelopment of the city's culture over the next few decades. Components of the plan include rebuilding artistic talent pools, repairing and rebuilding cultural venues, teaching the city's cultural traditions to young people, attracting national and international investment, requiring shared public and private investment in the city's infrastructure, and "marketing

New Orleans' Racial Composition

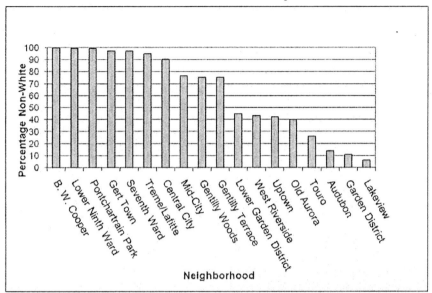

Table 3.1. Data from The Brookings Institute (2005). *New Orleans after the Storm: Lessons From the Past, a Plan for the Future.* Washington, DC: Brookings Institution.

New Orleans as a world-class cultural capital" (City of New Orleans, 2006, p. 14). By adhering to the plan proposed by the mayor's office, many leaders believe that:

> The proposed public and private investments will revive the City's cultural base, benefit businesses and residents of every neighborhood, endure the return of displaced artists and cultural workers, restore leading cultural facilities and create new cultural venues that celebrate the City's unique musical history and the cultural traditions of its diverse neighborhoods, revitalize street life and performances, increase tourism, and lever other investments many times over. (City of New Orleans, 2006, p. 14)[20]

Since the storm, the loss of culture indicative to New Orleans is one of the primary concerns of the city's residents. Culture is significant to the survivors because it has shaped a sense of place and place attachment; however, with an emphasis on tourism and the outside workers that are bound to follow, how significant will the renewing of New Orleans culture be to them? There is no doubt that those material components and festive aspects of New Orleans culture will remain significant to the workers who will flock to the city for employment opportunities in the tourist industry; however, the city's economic revival rests in the commodification and selling of the "traditional New Orleans experience"— complete with jazz, food and fun—as opposed to its cultural value. Gothman (2002) explained that the "concept of commodification refers to the dominance of commodity exchange-value over use-value and implies the development of a

Average Household Incomes

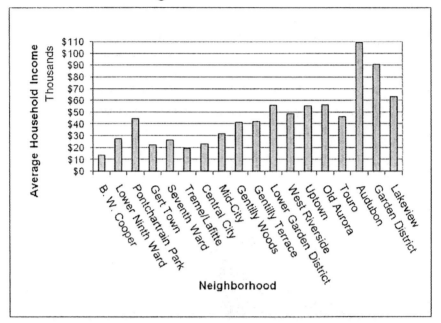

Table 3.2. Data from The Brookings Institute Brookings Institute (2005). *New Orleans after the Storm: Lessons from the Past, a Plan for the Future.* Washington, DC: Brookings Institution.

consumer society where market relations subsume and dominate social life" (p. 1737).

Therefore, in the context of selling the city's culture and sense of place, local customs, rituals, festivals, and ethnic arts will become tourist attractions performed on cue when tourists want to be entertained (Gothman, 2002), devaluing the traditional experiences in the eyes of traditional practitioners. Devaluation of traditional culture by residents will occur as traditional events become trivial and less significant in relation to annual events with cultural meaning; traditional events will give way to more financially motivated activities that can be "packaged" and sold as "culture on demand." This was already distinguishable among the city's residents prior to the storm with Mardi Gras.

> In the past, Mardi Gras developed as a relatively indigenous celebration for local residents that existed outside the logic of market exchange and capital circulation. Today, tourism entrepreneurs and urban boosters aggressively market Mardi Gras as part of a larger tourism-oriented strategy to encourage people to visit and spend money in the city—a tendency some local residents believe will result in the devaluation of the celebration for the city. (Gothman, 2002, p. 1744)

New Orleans' Poverty Rate

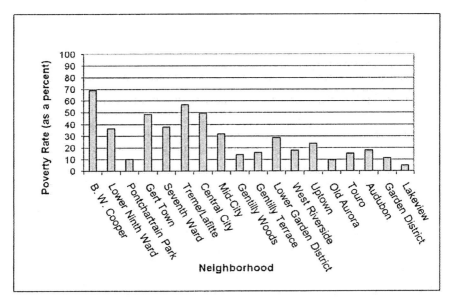

Table 3.3. Data from The Brookings Institute Brookings Institute (2005). *New Orleans after the Storm: Lessons from the Past, a Plan for the Future*. Washington, DC: Brookings Institution.

Local concerns about the establishment of a dominant tourist sector is exacerbated by a more commodified tourist industry that will be centered on less culturally significant activities rooted in and connected with the landscape.

In the pursuit of economic success, the city and private businesses will attempt to take as much economic advantage from the city's sense of place as possible; however, this may have a detrimental effect on the city's culture in the long-term. Because culture will be an economic stimulus, and possibly the most relied on sector in the region, business and city planners will have to avoid the occurrence of *Disneyfication*[21] of the urban and cultural environments. With the occurrence of *Disneyfication*, local traditions, famous buildings and landmarks, and other heritage sites and events become "hyper-real," causing people to lose the ability to distinguish between the "real" and the "illusion" (Gothman, 2002) or between original culture and marketing-manufactured culture.

Additionally, Harvey (1989) and Holcomb (1993) explain that with the expansion of the tourist industry into a dominating sector, place promotion no longer concerns itself with informing or promoting in the ordinary sense (i.e., selling the original cultural characteristics of New Orleans) but rather is geared toward manipulating desires and tastes through imagery that may or may not have anything to do with the product being sold (New Orleans' culture) (Harvey, 1989; Gothman, 2002). In essence, the tourist industry ceases to "sell" New Orleans' true culture in this circumstance and instead involves itself in adapting, reshaping, and manipulating images and sense of place to be desirable to targeted consumers (Gothman, 2002). Functioning within this type of economic and social environment, New Orleans' culture may inexplicably change, not due to alterations in just the demographics and physical topography of the area, but rather because New Orleans' culture will develop directly in conjunction with what is profitable to the tourist industry and what is not. Moreover, shifts toward tourist-dominated economics in an urban setting have historically coincided with population decline, white flight to the suburbs, racial segregation, poverty, and a host of other social problems, including crime, fiscal austerity, poor schools, and decaying infrastructure (Gothman, 2002), all of which were observable in New Orleans prior to Katrina. Therefore, in looking toward the future, one can foresee that although the expansion of the service sector may aid the economic redevelopment of New Orleans in the immediate short-term, in the long-term, social conditions may to revert back to what they were prior to Katrina in addition to losing the cultural significance, place attachment, and sense of place that many of the residents held for New Orleans prior to the storm.

Summary

The economic and cultural landscapes of New Orleans have profound effects on each other because they enable or constrain the development of the other. Moreover, the economic landscape sets limits on the ability of the city (and, on a macro-level, the state) to provide a variety of social services that would aid in increasing the quality of life for many of the city's past residents and make new migration to the city more desirable. Additionally, these conditions promote a specific lifestyle and cultural landscape indicative to the city of New Orleans due to residents' ability to live in this economic landscape. The cultural landscape, however, is not only a direct result of the economic landscape but is a significant influence on it as well. The cultural landscape, with its emphasis on root culture and festive traditions, stimulates the economic landscape through its production of services and products that people worldwide want to experience.

The effects of the two landscapes on each other cannot be underestimated and will always persist. In the future, one or both of these landscapes will change, whether due to demographic shifts or due to the dominance of a certain economic sector (i.e., reconstruction—*disaster capitalism*—and "*Disneyification*"), which will inevitably alter the other. In essence, the initial microeconomic, macroeconomic, and cultural effects of Hurricane Katrina will be deeper and more lasting (Dekaser, 2005) than previous hurricanes, such as Betsey and Camille. The question is not *if* either of these two landscapes will change, it is *how*. Although changes to the economic and cultural landscapes will occur, only through the manipulation of the political landscape will residents and officials be able to guide the changes to any effective degree to the benefit of everyone.

Notes

1. The first French possession of the colony occurred between 1699 and 1769, after which the Spanish took control between the years 1769 and 1803. Then, for a brief time

during the French Revolution, the Spanish gave back control of the colony to France in 1803. Later that year, France sold the Louisiana Territory to the United States.

2. Kephart (1948) explained the various distinctions between blacks with different percentages of African blood. Distinctions between the castes and the terms associated with their variations in ethnicity are as follows: (1) Pure Black—Negro, (2) three-quarters Black—Sambo or Mangro, (3) half Black—Mulatto, (4) one-quarter Black—Quadroon, and (5) less than one-quarter Black—Octoroon, Mustifee, or Pass for White.

3. Florence (1997) explained that before the close of the 19th century, Blacks and Whites were buried close to one another in New Orleans even though slavery was still in practice. The lack of segregation within New Orleans' cemeteries, which is seen later under American control, illuminated the close social and cultural ties among both races and the lack of emphasis placed on race, which can be attributed to the interracial breeding and social interaction that took place in the city (Miller & Rivera, 2006b).

4. The authors note that equal economic access between races was considerably lacking; however, freed-persons enjoyed a level of economic access in New Orleans prior to the Civil War that was high relative to other urban locations across the United States at the time.

5. Individuals of African decent whose ancestors had engaged in interracial mixing to the point where they would be mistaken as being White were called passé-blanc.

6. Jim Crow Laws were firmly embedded into American culture by the 1890s and "dictated where blacks could eat, which seats they could occupy in theaters and on buses and trains, which jobs they could perform, where they could live, which water fountain they could use, and which beaches and parks they could visit.... Beginning where statutory restrictions ended, Jim Crow customs and racial etiquette seized every opportunity to belittle and humiliate African Americans" (Chafe, Gavins, & Korstad, 2001, as cited in Dawson, 2006).

7. Prior to the Great Mississippi Flood of 1927, the area around New Orleans was economically viable due to its port location in conjunction with its investments in agriculture; however, due to poor governmental response to aiding Blacks in the area, many people decided to leave the area after the waters receded. According to Barry (1997), after the flood, Black sharecroppers decided to move north to Chicago and west to Los Angeles to find better work and escape social conditions imposed by local and federal government officials that were racist in nature. The Great Mississippi Flood of 1927 was a natural disaster similar to Katrina that resulted in the migration of thousands of Blacks to other parts of the country (Rivera & Miller, 2007a), which had profound detrimental effects on the New Orleans social and economic landscapes, in addition to the social and economic landscapes of wherever these migrants decided to settle.

8. By limiting where these people could live, the larger society also limited, to some degree, Black social interaction primarily to these segregated areas. Because they were limited to living in these areas, they may not have been able to acquire the aesthetic experiences that are indicative to other parts of the city or been able to experience other social interactions that took place in other parts of the city. These limitations, which were socially imposed, severely limited Black interactional potential in relation to White.

9. See Habermas (1992[1996]) for a full discussion of the lifeworld concept.

10. New Orleans' population prior to Katrina was approximately 475,000 people with 67 percent of the population being African American. Estimates indicate that the future population will only be 350,000 with only 35 percent to 40 percent of the population being African American (Pastor et al., 2006).

11. According to Lipsitz (1988), "On the surface, the core practices of the Mardi Gras Indians resemble quite conventional behaviors by other groups. Under the aegis of carnival, they form secret societies, wear flamboyant costumes, speak a specialized language, and celebrate a fictive past" (p. 102). Examples of New Orleans' Mardi Gras In-

dian Tribes or Krewes include the following: the Corps de Napoleon, The Ducks of Dixieland, the Knights of Babylon, the Krewe de C.R.A.P.S., Le Krewe d'Etat, Krede du Vieux, the Krewe of Alla, the Krewe of Amon Ra, the Krewe of Apollo de Layfayette, the Krewe of Armeinius, the Krewe of Bacchus, the Krewe of Barkus, the Krewe of Carrollton, the Krewe of Cork, the Krewe of Elvis Marching Club, the Krewe of Endymion, the Krewe of King Arthur, the Krewe of Mid-City, the Krewe of Morpheus, the Krewe of Muses, the Krewe of OAK Uptown Carnival Society, the Krewe of Orpheus, the Krewe of Pegasus, the Ponchartrain, the Krewe of Rex, the Krewe of Shangri-La, the Krewe of Trucks, the Krewe of Zulu, the Lords of Leather, the Mardi Gras Indians of Mardi Gras, and the Mystic Krewe of Satyricon.

12. It is estimated that 110,000 of New Orleans' 187,000 homes were flooded, of which 90,000 sat under more than six feet of water for days. It is estimated that somewhere between 30,000 and 50,000 homes city-wide will have to be demolished or need extensive repairs (Pastor et al., 2006).

13. In October 2005, a total of 281,745 Louisiana residents filed for unemployment benefits, all of which cited Katrina as the cause of their joblessness. In New Orleans specifically, 47 percent of all workers in the seven parishes that compose the city filed for unemployment for the same reasons (Pastor et al., 2006; U.S. Bureau of Labor Statistics, 2005).

14. A disaster economy is characterized by an immediate boom of major prosperity due in part to federal, state, and local insurance payments, which attempts to foster the reestablishment of economic forces.

15. Immigrant presence in the Katrina-affected areas has steadily increased between 1990 and 2000. In Louisiana alone, the Hispanic population increased by 15.8 percent. More specifically, about 15,000 Latinos officially resided in the New Orleans metropolitan area in 2000; however, that number is believed to be much higher due to the large influx of Latinos, particularly Hondurans, into the area over the last several years (Muniz, 2006).

16. Although there are several estimates on the size of the Latino population, a precise number is almost impossible to estimate considering that many Latinos who live in the South are foreign born and undocumented (Dyson, 2006).

17. The increased Latino presence is already affecting the cultural landscape of the city, mostly manifested in food. Although Latin cultural characteristics are hard to find outside of construction areas (Cowen, 2006), these same characteristics will soon diffuse to other areas of the city, specifically where this population resides and wherever else the market for these types of food take them.

18. See also Gold & Ward, 1994, Gottdiener, 1997, Reichl, 1997; Short, 1999; Strom, 1999; and Zukin, 1996.

19. For more details about the rebuilding strategy, see City of New Orleans (2006).

20. For more information, see City of New Orleans (2006).

21. Gothman (2002) explained "the transformation of public spaces into privatized 'consumption' spaces and the latest attempts by tourism entrepreneurs and other economic elites to provide a package of shopping, dining, and entertainment within a themed and controlled environment—a development that scholars have called the 'Disneyfication' of urban space" (p. 1738). For more information, see Eeckhout, 2001, Reichl, 1999, and Sorkin, 1992.

Chapter 4
The Political Landscape

Natural disasters occur in a political space. They are not driven by politics but nor are they immune from politics.

—Cohen & Werker (2004, p. 3)

The political landscape of an entire nation is not a uniform single landscape, rather a composition of unique political landscapes within its geographic boundaries. Just as individual personal experiences influence a person's degree of place attachment and sense of place, all social, economic, and political decisions also influence one's sense of place. Public policy within the political landscape of any location directly affects the lived experiences and has a direct impact on a person's sense of place. Moreover, the political landscape of any geographic location is shaped by and, at the same time, shapes the socio-economic and cultural landscapes. When catastrophic natural disasters take place, the political landscape changes; however, the trajectory of change is not predictable. This chapter outlines the political landscape of New Orleans prior to Katrina and documents its on-going changes. By understanding the political landscape of New Orleans, the United States' political context prior to Katrina, and the political landscape evolving in New Orleans following Katrina, one can come to know the effects of the political landscape on the residents' sense of place as a direct result of politics and public policy.

Politics of Protection: New Orleans before Katrina

In an effort to relieve the subsequent effects of flooding on the city and the greater Mississippi Delta, early city leaders enacted levee construction policies. The earliest man-made levees in the New Orleans area were constructed within the first few generations of the city's founding. Early levees were constructed through the patronage of community businessmen and local governments, which produced approximately 135 miles of levees relatively early in the region's history (Congleton, 2006). The construction of these early levees did have some benefit, but they did not stop the occurrence of flooding in the area. Also, their construction perpetuated the occurrence of flooding on the banks of the Mississippi opposite from where they were built. In an attempt to ensure commerce, levee building policies sparked an unending cycle of building and rebuilding that shaped the land and encouraged more levee construction.

> levee building never stopped; levees were extended above and below New Orleans, then to the opposite bank. Those levees increased the pressure on old ones. The reason is simple: when the river was leveed on only one bank, in flood it simply overflowed the opposite bank. But with both banks leveed, the river could not spread out; therefore, it rose up. Thus the levees, by holding the water in, forced the river higher. In turn, men tried to contain the flood height by building levees still higher. (Barry, 1997, p. 40)

The constant building of levees created external problems for residents who lived both up and down the river from the levee sites, which manifested themselves downriver as higher waters, stronger currents, and flooding due to decreased flow rates that would have been higher if the river was allowed access into formerly unrestricted flood plains. Because levees aided both in controlling the floods and encouraging flooding in other areas, the only apparent way to deal with the situation was to continually build "bigger and better" levees over time (Congleton, 2006).

The federal government supported the construction of levees as a mechanism for flood mitigation with The Swamp Land Acts of 1849 and 1850. The Swamp Land Acts allowed states to use internal revenues to construct levees and drain channels, thereby officially placing the responsibility of flood mitigation in the hands of local government authorities. Moreover, in 1885, the Mississippi River Commission adopted a *levees only* policy,[1] which they believed would be the most successful way of controlling flood waters (Rivera & Miller, 2006). This policy, motivated by idealistic engineering theory, argued:

> that alluvial rivers, like the Mississippi, always carried the maximum amount of sediment possible and that the faster the current, the more sediment the river *had* to carry. His [the engineer's] hypothesis further argued that increasing the volume of water in the river also increased the velocity of the current, thus compelling the river to *pick up* more sediment. The main source of this sediment had to be the riverbed, so confining the river and increasing the current forced a scouring and deepening of the bottom. In effect, adherents to this theory argued, levees would transform the river into a machine that dredges its own bottom, thus allowing it to carry more water without overflowing. (Barry, 1997, p. 41)

However theoretically plausible it was, the policy did not work in reality. Because the policy failed to create a plan that would control flooding (whether due to problems related with implementation or because the policy itself was flawed), President Coolidge signed the Flood Control Act of 1928 (PL 70-391) after the Mississippi Flood of 1927,[2] which ended the predominant use of the levees only policy (Rivera & Miller, 2006). This by no means ended the use of levees as control mechanisms, but this legislation enabled flood control to incorporate the construction of "bigger and better" levees and modest efforts to restore flood plains and spillways to divert flood waters away from New Orleans (Congleton, 2006).

Historically, flooding in the area can be attributed to the Mississippi River and Lake Pontchartrain. Congleton (2006) explained that when the lake overflowed and caused flooding like the Mississippi did, local officials attempted to control nature by constructing "bigger and better" levees and spillways into and around the lake, such as the Bonnet Carre Spillway. Constructed in 1932, this spillway is six miles long and was built to divert flood waters from the Mississippi into Lake Pontchartrain to take pressure off river levees when waters were excessively high; in essence, it acted as a "relief valve" for the river. Since its construction, the spillway has been used eight times as of 1997 (U.S. Department of the Interior, 1997), each time increasing the risk that the lake would be the cause of flood damage (Congleton, 2006). Moreover, as explained in Chapter 2, the man-made efforts to drain the Mississippi into Lake Pontchartrain and, subsequently, Lake Pontchartrain into the Gulf of Mexico actually served to direct the storm surge from the Gulf into New Orleans when Katrina struck.

Without the fiscal support of Congress, responsibility of generating funds for further levee construction fell to local levees boards and other local government authorities. When left to local officials, major regional responsibilities can be problematic, and support for such issues can be hindered by local initiatives. When faced with a lack of funds to invest in capital projects, new monies have to be acquired from new sources. It is ironic that in the pursuit of finding funds to construct new levees, the local and federal governments were supporting development programs that placed anyone residing in the newly constructed areas at a higher risk from flooding.

In 1965, flooding caused by Hurricane Betsy prompted Congress to authorize the creation of the Lake Pontchartrain and Vicinity, Louisiana, Hurricane Protection Project. This project was initially inspired by its dual advantages, which were to protect all of Orleans Parish and the northern portion of Jefferson Parish from storm surge flooding and category 3 hurricanes as well as to facilitate continued urbanization of the hazard prone area (Burby, 2006). The fact that the project would protect preexisting development was eclipsed by the potential to bring more development into the area.

> In fact, protection of *existing development* accounted for only 21 percent of the benefits needed to justify the project. An extraordinary 79 percent were to come from new development that would now be feasible with the added protection provided by the improved levee system. (Comptroller of the Currency, 1976, as cited in Burby, 2006, pp. 174-175)

Support from federal and local government officials for this course of action was predicated on an expanded tax base in the event the project was successful. According to Burby et al. (1999), this seemingly inventive strategy is commonly exercised by both the federal and local governments. By increasing residential and business exposure to risks, individual stakeholders will work to seek alternative methods of avoiding hazards. Intensive development fosters increased exposure to risk in specific areas while sometimes stalling local interests in addressing adequate hazard planning that include more progressive land-use strategies.

> When hazardous areas are viewed by landowners and developers as reasonably safe, profitable places for development, land use approaches to hazard mitiga-

tion can be viewed by economic interests and local governments pursing economic growth as a threat to be avoided rather than a good to be fostered. (Burby et al., p. 249-250)

Moreover, Burby (2006) contends that as a direct result of Congress's authorization of the Lake Pontchartrain Hurricane Protection Project, Jefferson Parish added 47,000 housing units and Orleans Parish added 29,000. Lewis (2003, as cited in Burby, 2006, p. 175) noted that the New Orleans' metropolitan area "exploded into the swamps" surrounding the city on land that caused the newly developed area to sink at various rates, which further increased vulnerability.

Another development program that was initially federally supported dealt with land used by NASA for a Michound rocket assembly facility. This facility was a major source of employment in the region and, therefore, required housing for its workers. Although federal support ended in 1975, the development projects were renamed Orlandia and New Orleans East and were continued as private ventures that would support the housing of 250,000 residents (Burby, 2006). Like the Lake Pontchartrain Hurricane Protection Project, these development projects went forward due to similar perspectives held by local planners, building inspectors, public works engineers, and residents regarding their awareness of natural hazards and their low priority (Berke, 1998). Groups that are responsible for planning, developing, and residing in these relatively high-risk areas perceive hazards as unavoidable.

> Respondents [local planners, building inspectors, public works engineers, and residents] consistently view natural hazards, especially long-shot ones posed by low-probability/high-consequence events, as facts of life and acts of nature that are often inexplicable and completely unavoidable. The importance of preparing for a disaster in the distant future and risk-averse action is likely to be eclipsed by more immediate and pressing concerns (street potholes, waste disposal, and crime) that affect people almost daily. (Berke & Campanella, 2006, pp. 194-195)

Development of marshes and wetlands continued with a focus on solving issues regarding the expansion of the local tax base and increasing the available workforce that would expand the local economy. Between 1970 and 2000, more than 22,000 new housing units were built and even more were planned (Burby, 2006). In the *1999 New Century New Orleans Land Use Plan*, the city appeared adamant about future construction and development into the highly flood prone area:

> Moreover, there are extensive opportunities for future development of the vacant parcels that range from single vacant lots to multi-thousand acre tracts. Long-term, these development opportunities represent not only population increases but also significant potential employment for the city. (City Planning Commission, 1999, as cited in Burby, 2006, p. 176)

Even with the political and fiscal investments made into these developments and the protection offered by the "bigger better levees," Katrina placed all of these development projects underwater in defiance of developers and the Army Corps of Engineers. The manmade levees, created to defend the city from flooding, gradually caused the city to sink and exposed the city to even greater risks in the event the levee system failed; the creation of "bigger better levees" directly contributed to the increased flooding risks associated with category 4 and 5 hurricanes on the city because any levee failure would allow more water into the city and make flooding worse (Congleton, 2006).[3]

Politics of Protection and Increasing Vulnerability

The actions taken by local government authorities regarding flood mitigation throughout most of the history of New Orleans have increasingly subjected residents to an ever-growing vulnerability to natural disasters. Cutter (1996) defined vulnerability as the potential for losses that could occur to an individual. Vulnerability varies from person to person and place to place. However, in the case of New Orleans, vulnerability was not only experienced on a micro level (i.e., the individual) but also on a macro level with regard to social groups. Cutter and Emrich (2006) defined social vulnerability as the following:

> the susceptibility of social groups to the impacts of hazards, as well as their resiliency, or ability to adequately recover from them. This susceptibility is not only a function of the demographic characteristics of the population (age, gender, wealth, etc.) but also more complex constructs such as health care provision, social capital, and access to lifelines (e.g., emergency response personnel, goods, services). (p. 103)[4]

In this framework, a group's socio-economic characteristics directly affect their level of vulnerability to natural disasters because of their limited ability to cope with the effects of a detrimental hazard. As discussed in Chapter 3, the socio-economic status of many people affected by Hurricane Katrina directly impinges on their ability to return.[5]

Although the social characteristics of groups are factors affecting their vulnerability, one factor that cannot be ignored is place vulnerability. One of the two primary components of place vulnerability are "those factors of the environment that lead to increased potential for hazardous events to occur, or physical vulnerability (e.g., Do you live in a hurricane area, are you near a chemical or nuclear facility?)" (Cutter & Emrich, 2006, p. 106).[6] Place vulnerability can also be described in relation to variables correlated with exposure, such as proximity to the source of threat, incident frequency or probability, magnitude, duration, or special impact (Cutter, 1996). Urban development into areas prone to hazard reoccurrence significantly increases the vulnerability of residents, which seems to be something the federal and local governments have chosen to ignore in the New Orleans region.

The choice of local governments to allow development in topological regions prone to recurrent disasters is perpetuated by the creation of federal flood insurance programs that work to compensate many businesses and homeowners for losses incurred by catastrophe. Locally, governments tend to provide the minimal amount of protection possible to both attract residents and qualify for federal insurance policies. New Orleans was not an exception to this practice. Burby (2006) noted that the Orleans Parish Levee Board lobbied the Corps of

Engineers to only protect the city against events that were expected to occur every 100 years as opposed to every 200 years (i.e., a category 5 hurricane).

In the 1980s, issues associated with reliance on federal insurance programs surfaced when the Federal Insurance Administration (FIA) brought a subrogation suit against the Jefferson, Orleans, and St. Bernard Parishes.[7] The FIA's suit centered on the notion that it had paid excessive flood insurance claims due to the parishes' failure to adequately maintain levees and enforce elevation requirements of new construction, which directly resulted in the flooding of buildings whose owners applied for compensation from the federal flood insurance program. The court ruled in favor of the FIA and ordered the parishes to improve levee maintenance and enforce construction requirements (Malone, 1990). Although the FIA has had the opportunity to file lawsuits against negligent governmental units, this has not been a motivating factor in getting governmental units to employ effective mitigation projects. Insurance agencies have attempted to overcome the moral hazards associated with federal insurance programs, but it appears that local governments operate as if the FIA may overlook these moral hazards.[8]

The obvious question that presents itself is "Why, if insurance providers are reluctant to provide policies to reoccurring high-risk clients, do local governments continue to invest in development projects that the insurance companies balk at when asked to fulfill claims?" This occurs because local governmental units have been given the impression that although they are partly liable for their approval to build in high-risk areas, it is traditionally the federal government's responsibility and practice to financially compensate victims. This ideology has developed in response to the Disaster Mitigation Act of 2000 and the National Flood Insurance Reform Acts of 1994 and 2004 (Burby, 2006). Although the Disaster Mitigation Act of 2000 set in place standards and policies regarding the financial burden of mitigation plans, local and state governments have been hesitant to implement any changes to their preexisting mitigation strategies. Additionally, although the Disaster Mitigation Act of 2000 and the National Flood Insurance Reform Acts of 1994 and 2004 stipulate that mitigation practices must be maintained and that there should be more comprehensive planning in reference to disaster effects in a given development area, the acts also acknowledge that the federal government will take financial responsibility in the event local mitigation efforts fail to dispel destruction of property.

> To minimize the adverse financial consequences for individuals and businesses when steps to make development safe from hazards fail (known technically as residual risk), the federal government has provided generous disaster relief, particularly for homeowners; low-cost loans to ease business recovery; income tax deductions for uninsured disaster losses; and subsidized flood insurance. (Burby, 2006, pp. 173-174)

Consequently, these laws, which were created to reduce development in areas prone to risk, have instead stimulated development into hazard prone areas, increasing vulnerability to natural disasters.[9] A research design created by Cutter, Boruff, and Shirley (2003) and used by Cutter and Emrich (2006) supported the notion of increased vulnerability over time that can be attributed to urban development and the socio-economic factors of New Orleans:

> The higher social vulnerability score in 2000 compared to 1960 suggests an increase in social vulnerability over time, unlike the other counties in the affected region. This indicates that not only do persons living in New Orleans Parish generally have less ability to cope with major natural disasters than their counterparts in the other parishes, but they also have less ability to rebound from catastrophe than they did in 1960. (Cutter & Emrich, 2006, p. 108)

Indirect federal support for development into hazard prone areas has also had a bureaucratic effect on local government leaders and committees. Because the federal government has undercut incentives to dramatically invest in mitigation practices through federal compensation, bureaucrats and local public officials have developed an "It's not really my problem" or an "I'll let the next guy deal with it" mentality. Shughart (2006) explained that these perspectives are motivated by the goals of reelection or reappointment to office that may be compromised if an administrator pushes for extensive investment into programs or policies that have little tangible and immediate benefits. Therefore, the public sector assigns less weight to the future benefits and costs to a policy or program of action that may be realized in another term of office.

Additionally, because the infrastructure deteriorates slowly and unnoticeably, underinvestment occurs in the form of maintenance, repairs, and new plans for coping with natural disasters. In reference to Hurricane Katrina and New Orleans, there were four separate levee boards responsible for the maintenance and repair of the levees systems around the city (Shughart, 2006). Due to the board members' reliance on political patronage for reappointment,[10] it was a detriment to their future careers to create situations that were politically volatile, such as raising taxes or rezoning specific areas and making them unsuitable for residence, so they instead promoted situations that were politically advantageous, such as developing land to increase employment and the tax base even though the development was potentially hazardous.

> Over time, using powers of eminent domain for flood-control projects, the board of the Orleans Levee District became the largest landlord at Lake Pontchartrain. It built two marinas there; constructed parks, walking paths, and other amenities along the lakefront levees; and in order to spur development at its marinas, built roads, a commuter airport, and a dock it leased to the Belle of Orleans, a floating casino, in return for a cut of the gaming revenue. . . . The humdrum, largely invisible job of levee maintenance took a backseat to more newsworthy—and more politically rewarding—lakefront development initiatives. (Shughart, 2006, pp. 35-36)

Seemingly, local governments tend to follow courses of action that have tangible rewards for their administrations. Even though their decisions may have an adverse effect on the relative safety of their constituents, their decisions are indirectly supported by the presence of federal programs that compensate businesses and homeowners for exposure and losses due to natural hazards.[11] Even though development projects seem to be often supported by the federal government as a whole, these practices have made emergency managers hesitant to release resources to aid-asking municipalities that the Federal Emergency Management Agency (FEMA) sees as corrupt for contributing to the phenomena of moral hazards.

> Likely exacerbating FEMA's over-cautiousness further was its reluctance to trust local officials due to the widely-held perception of rampant public-sector corruption in New Orleans (and the state of Louisiana). Unable to determine the credibility of the mayor's and governor's claims of need, of instance, FEMA might have felt compelled to wait and gather more information about the accuracy of these claims before acting. It might have also led FEMA to mistrust local officials' ability to be good stewards of federal (an even their own) disaster relief resources. In this way, state and local public sector corruption may have contributed to FEMA's delayed response. (Sobel & Leeson, 2006, p. 59)

Therefore, because of the city's participation in ventures that were risk generating and solely politically beneficial to city officials, FEMA was justified in its wariness to give New Orleans the support it needed without questioning the situation thoroughly.

Accountability and Who is to Blame

The people's ability to hold the government accountable for its actions, or lack thereof, is one of the hallmarks of democracy. In America, blame regularly alternates among the federal bureaucracy, the presidency, the states, and the local governments.[12] In most cases, citizens expect the government to perform certain duties, such as provide protection, which can be defined in any number of ways; however, in the case of disaster mitigation, accountability is not so simple. According to Freudenburg (1993), when a failure to perform specific behaviors occurs, those citizens assessing the institutional actors are less likely to depend on these specialized individuals representing the institutional actors, thus rendering the individual and institutional actors less credible. It is difficult to separate trust, or lack of trust, in an individual from a specific institution; rather, the citizens of a community are likely to blame the entire structure that encompasses trust, agency, responsibility, or other expected obligations of the collectivity. In situations where individuals are the heads of institutions, the general public is inclined to blame that individual. In such situations, individuals can be terminated or they may resign from their positions to regain some trust in the institution or agency on the behalf of the general public. However, when there is not a specific face to an institution regaining the general public's trust in the institution is severely limited. Regardless of who or what institution is held responsible, the public's faith in the government has been shaken when explanations are not met:

> As Jean-Jacques Rousseau (1800) argued in 1763, one of the glues that holds a people together is the contract between a government and the people—a contract that stipulates that a government is for the goodwill of the people, that it will protect and aid the people, and that the people will give over decision making and sovereignty to that government. In the United States, we assume that if disaster strikes, we have a social contract to which a responsible government will abide. (Ethridge, 2006, p. 802)

In an attempt to gain and keep the trust of the general public, Congress passed the Government Performance and Results Act in 1993; the law demands that federal agencies develop and implement strategic plans with performance measures and reports to be used by the executive and legislative branches of government in making budgetary decisions (Gormley & Balla, 2004); furthermore, as a result of this legislation, those agencies that performed poorly would receive less funding whereas those that performed well would receive incremental increases to their budgets. Also, agency reports and performance measures would aid policymakers in redesigning failing programs as well as provide information on the best practices. Although this plan held agencies fiscally responsible for their actions, it was extremely difficult to implement the policy, especially in regard to agencies with significant intergovernmental relationships (Gormley & Balla, 2004).[13] The policy may have been successful, but the true result of the policy was never realized. According to Gormley and Balla (2004), even after the passage of the Government Performance and Results Act, legislators tended to ignore performance data, even from the agencies that provided informative and complete reports, and make congressional appropriations just as they had before the passage of the law.

Blame and the Many Contributors to the Political Landscape

The political landscape of New Orleans is shaped by local and national organizational, institutional, and personal influences. Some of the influences come from specific people who create and augment its structure, regardless of the passage of natural events; however, some are specifically linked to the development of the landscape in the event a natural disaster occurs. For the purposes of explaining the possible alterations to the political landscape, the roles of the four actors in this landscape (the federal bureaucracy, the president, the state of Louisiana, and the city of New Orleans) in reference to mitigation, preparedness, relief, and response are detailed. These four actors were chosen because of the widespread blame placed on them by the media and the public for their inefficient response, but they are by no means the only actors that influence the political landscape. For example, local and national nonprofit organizations, private business, advocates, lobbyists, and international nongovernmental organizations have profound influences on any political landscape. For the purpose of this discussion, we limit the conversation to the New Orleans political landscape and concentrate on the four actors previously identified.

The Political Landscape and Bureaucracy

In the aftermath of Katrina, there were no agencies in government that readily accepted responsibility for the slow response and relief efforts. According to Waugh (2006), leaders at all levels of government appeared "disconnected at best and insensitive and incompetent at worst" (p. 11). From the American public's point of view, FEMA was to bear the bulk of responsibility; however, this over-simplified blame was accorded to the agency that had been primarily responsible for dealing with disasters in the past. Moreover, public sentiment toward FEMA, although not entirely misplaced, was misled due to the agency's mismanagement and reorganization, which severely limited the agency's ability to implement any strategies in the event of a disaster, but this is a fact little known to the general public. Prior to 2002, FEMA had operated independently of any other department or agency. During the Clinton Administration, FEMA's director was given de facto cabinet status for nearly 8 years (Haddow & Bullock, 2003; Sylves, 2006); however, when Congress created the Department of Homeland Security (DHS), FEMA lost its cabinet status and was incorporated into a mix of 18 other agencies, all competing for pieces of the same budget.[14]

Prior to its incorporation into the DHS, FEMA was responsible for dealing with all types of catastrophic incidents.[15] However, in day-to-day practice, FEMA, not considered a first-responder agency, most frequently dealt with natural disasters depending on local first-responders. According to Platt and Rubin (1999), FEMA was the only component of the DHS specifically responsible for reducing the losses associated with non-terrorism related disasters. Although still responsible for non-terrorist related disasters, the agency's incorporation into DHS was problematic and the DHS's organizational situation when Katrina struck made the situation worse.

> DHS is a mammoth, complex, and organizationally diffuse federal bureaucracy that was less than two years old when Katrina struck. Since 2003, DHS first headed by former governor Tom Ridge and then by attorney Michael Chertoff, has been put through many reorganizations, president-sanctioned, that have in some ways compromised its ability to manage large-scale, multistate disasters. (Sylves, 2006, p. 30)

With FEMA's new structural position in the DHS, the agency no longer reported to the president directly but instead reported directly to DHS management. This new reporting structure relegated this relatively small agency to just "another of many" separate departments within a massive bureaucracy, causing it to be overshadowed by much larger and better funded entities within the DHS (Tierney, 2005). As FEMA's budget shrank, personnel and resources were shifted to counterterrorism programs, detrimentally affecting not only the number but also the morale of remaining FEMA personnel (CNN, 2005).

> The agency [FEMA] is effectively being dismantled with its constituent parts being moved to other parts of DHS. DHS itself is focused on preventing terrorist attacks and is not organized to deal with natural disasters or even terrorist-caused disasters. (Waugh, 2006, p. 17)[16]

The reorganization of FEMA into the DHS required a reassignment of leadership and duty responsibilities, as assigned by the National Response Plan[17] and the National Incident Management System.[18] With the restructuring of the DHS, natural disaster emergency management became secondary to law enforcement, port security, intelligence, border control, immigration, and transportation security (Sylves, 2006). Although all these changes were made with the Bush Administration's approval, some administration experts recommended that FEMA be removed from the DHS to restore its capabilities to deal with natural disasters

(Waugh, 2006);[19] however, as history has shown, the recommendations fell on deaf ears. Moreover, from a process standpoint, the incorporation of FEMA into a multilayered bureaucracy (which is inherent in centralized decision making) delayed the decision-making processes (Sobel & Leeson, 2006). Due to this sluggishness, the most notable success stories of the relief effort following Katrina came from those who ignored FEMA and took action without approval.[20]

The Political Landscape and Federal Government

Blame for the response efforts after Hurricane Katrina was also pointed at President George W. Bush and his political appointments. The reorganization of FEMA into the DHS forced veteran emergency managers in FEMA's prior offices of response, recovery, and preparedness into consulting or state emergency manager jobs (Sylves, 2006). The lack of qualified leadership within the residual agency allowed the president to take the opportunity of appointing people to those vacant positions.[21] Although patronage appointments are not a new phenomenon in politics, the appointment of seemingly inexperienced personnel into leadership roles, where emergency experience was required, aided in the government's inability to effectively respond (Sylves, 2006).

Bush was also blamed by the public and the media for the ineffective relief effort due to his seemingly uninterested and lethargic personal response. As noted by Dyson (2006), a great deal of blame fell on the Office of the President because it was correctly perceived that Bush could have, at anytime, ordered a more effective and time-efficient response; moreover, many of the affected population contended that Bush did not care about the concerns of impoverished Blacks. The President's insensitivity and delayed concern was hard to dispute considering his actions throughout the initial phases of the disaster.

> Nearing the end of a five-week-long vacation at his Crawford, Texas, ranch (Thomas et al., 2005), the president himself, seemingly unaware of the magnitude of the disaster on the Gulf Coast, continued to focus on other matters. On the day Katrina made landfall, he kept two previously scheduled speaking engagements at senior centers, one in Arizona, the other in California, to promote the new Medicare drug benefit program (Katrina Timeline, 2005). The next day, Tuesday, 30 August, with the *USS Ronald Reagan* as a backdrop, President Bush spoke on the Iraq war at a naval base in San Diego, California and then returned to Crawford for the final night of his vacation (ibid.). On the way back to Washington Wednesday, Air Force One flew over New Orleans for 35 min. to give him a bird's eye view of the conditions on the ground. (Katrina: What Happened When, 2005, as cited in Shughart, 2006, p. 38).

The lack of leadership initiative on behalf of the White House led to the eventual release of several statements, all of which attempted not to pass blame on to the other branches of government or to other government agencies. Although the White House did not blame anyone specifically, it did not directly take responsibility for the slow action taken by the federal government. Instead, the White House chose to report that all responsible entities for disaster relief and response acted correctly but were inadequate to the task. Simultaneously, the White House revealed its position on the duty of the military in similar future situations.

> The Government will learn the lessons of Hurricane Katrina. The response of government at all levels was not equal to the magnitude of Katrina's destruction. Many first responders performed skillfully under the worst conditions, but the coordination at all levels was inadequate. Four years after September 11th, Americans expect better. President Bush takes responsibility for the Federal government's problems, and for its solutions. It is now clear that a challenge on this scale requires greater Federal authority and a broader role for the U.S. Armed Forces—the institution most capable of massive logistical operations on a moment's notice. The President has ordered every Cabinet secretary to conduct a review of the response, and the President will make every necessary change to fully prepare for any challenge of nature or act of evil that could threaten Americans. (Press Secretary, Office of the White House, 2005)

Although the President took responsibility for the inefficient response efforts, the idea that he did everything he could in his position to aid the victims of Katrina was inaccurate. According to the National Response Plan, the DHS and the President had the authority to bypass the regular process of federal response to disasters if the situation predicated such an action; however, for unknown reasons, neither chose to act decisively (U.S. General Accountability Office, 2004, 2005).

The President chose to use the policy window opened by Katrina to enhance the use of the military in domestic situations. Additionally, to add emphasis on the need for the military in similar situations, the White House used language that placed natural disasters and terrorism on the same ideological plane, such as with the use of the phrase "any challenge of nature or act of evil." Although almost benign in appearance, by using language linking the evil of terrorism to the destruction caused by severe natural hazards, the President has generated a response from the public and Congress out of the fear of evil to support policy changes. It is interesting to note that the President would propose such a policy initiative after his lack of action, but it was most evidently a political decision. Shughart (2006) contended that Bush's lack of action during the crisis and his policy initiatives afterwards were the result of a lack of political incentive to secure votes. Moreover, Shughart (2006) argues that because Bush was in his second term and could not be reelected and the region affected by the disaster had consistently voted Republican since the Reagan years, it was possible that Bush believed that any action, or lack thereof, would not significantly alter voting patterns in the future. This explanation is extremely plausible considering his actions when he was attempting to be reelected for his second term.[22]

The Political Landscape and State Government

Many people, both above and below the state level of government, blamed the inefficient response effort on the governor. According to the U.S. House of Representatives (2006), the governor of every state is charged with ensuring the safety and welfare of its citizens. Included in this charge is the responsibility for emergency management, which requires the governor to coordinate state resources in an effort to prevent, prepare for, and respond to incidents such as natural disasters. Also included in the governor's executive responsibilities is the ability of the governor to order and enforce the evacuation of residents in the event of disaster and other emergency situations (Bea, 2005a; U.S. House of Representatives, 2006). The governor, as the commander-and-chief of his or her state's National Guard, has the authority to order troops to perform normal disaster relief tasks, such as search and rescue, debris removal, and law enforcement.

At the state level, emergency management agencies have the primary responsibility of preparing the state's disaster mitigation, preparedness, response, and recovery activities. In turn, these agencies coordinate with other state agencies and local emergency response departments to plan for and respond to potential or imminent disasters or emergencies. In Louisiana, it is required that the state's Office of Homeland Security and Emergency Preparedness determine the requirements for food, clothing, and other necessities and procure and preposition these supplies in the event of an emergency (Bea, 2005a, 2005b). Moreover, in the event of an emergency, the Louisiana Office of Homeland Security and Emergency Preparedness grants emergency powers to the governor and the parish and municipal governments in an effort to streamline the establishment of a Louisiana Emergency Management Plan (Bea, 2005b).

According to the U.S. Department of Homeland Security (2004), Governor Kathleen Blanco was personally responsible for her state's rescue and relief efforts and was authorized to take steps necessary in the pursuit of protection. On Friday, August 26, Governor Blanco declared Louisiana in a state of emergency, putting the state's disaster plan in action. Blanco immediately wrote President Bush, requesting that he issue a federal emergency declaration for Louisiana (Blanco, 2005; U.S. House of Representatives, 2006). Governor Blanco then handed over the decision to evacuate New Orleans to Mayor Nagin (Experts from Brown Hearing, 2005); however, the evacuation of Southeast Louisiana did not take place until late Saturday afternoon. During the evacuation phase,[23] the governor was personally involved in monitoring the contraflow traffic between Louisiana and Texas and between Louisiana and Mississippi. Audio recordings of Hurricane Katrina conference calls that were used as testimony before the U.S. House of Representatives state that by late Sunday afternoon, contraflow was halted, but residents could still evacuate on outbound highway lanes. The evacuation aided approximately 1.2 million Louisiana residents in reaching safety in their own vehicles, but thousands of residents without their own transportation, and those without an inclination to leave, were left to weather the storm in New Orleans (U.S. House of Representatives, 2006).

On August 26, in addition to the evacuation orders, Governor Blanco also authorized the mobilization of 2,000 Louisiana guardsmen, and the following day, Adjunct General Bennett Landreneau, Head of the Louisiana National Guard, called an additional 2,000 troops into active duty (National Guard, 2005; U.S. House of Representatives, 2006). Although the order to activate and deploy troops came before the storm, their deployment and the deployment of the state police were not as forthcoming as the order (Waugh, 2006). Waugh (2006) contended that the slow deployment of troops was not necessarily the fault of Governor Blanco or Lieutenant Governor Mitch Landreneau, but rather it was due to the lack of National Guardsmen domestically available because of assignments in Iraq and Afghanistan, which severely limited available personnel and equipment resources. Moreover, it seems that the communication infrastructure needed to withstand the destruction of a category 5 hurricane was not in place. According to Carrns (2005), Katrina disrupted communications systems, preventing Governor Blanco from gathering information from field officials, which is why she was not able to make a specific list of needs to Washington before Thursday, September 1. The disruption to communications and, as a result, the lack of up-to-date intelligence had a profound effect on the response and relief process at the state and federal levels.

> While Governor Blanco of Louisiana requested federal assistance well before the storm and the levee breaches, assistance was days away. Once the need to act was realized, the federal agencies were slow to deliver aid to victims stranded in New Orleans and other communities, slow to rescue those trapped in homes and hospitals, slow to recover bodies, and slow to deliver trailers to [the] disaster area...Hundreds of volunteer emergency response and medical personnel were waiting in frustration to be deployed by federal officials. (Waugh, 2006, p. 14)

Even though the state, represented by the governor, seemed to follow standard procedure when it came to dealing with Katrina, it appears that the state's preparedness efforts, in effect long before the storm, were not adequate.

The Political Landscape and City Government

The local government was blamed for the ineffective response and relief efforts following Hurricane Katrina. City governments bear the initial responsibility for disaster response and relief. As such, the mayor is responsible for initiating, executing, and directing operations during the advent of an emergency or a disaster affecting their constituent municipality. Included in his responsibilities, Mayor Nagin had the authority to order the evacuation of residents threatened by an approaching hurricane, just as the governor did (U.S. Department of Homeland Security, 2004). One of the main reasons for developing a local Comprehensive Emergency Management Plan is to ensure that a city has all the available resources needed to safely evacuate its threatened population. Moreover, the plan also stipulates that the city will make certain arrangements to address the immobile segments of the population.

> The plan also directs "[s]pecial arrangements will be made to evacuate persons unable to transport themselves or who require specific life saving assistance. Additional personnel will be recruited to assist in evacuation procedures as needed." (Office of Emergency Preparedness, 2004, as cited in U.S. House of Representatives, 2006, p. 50)

In addition to the city's responsibilities for evacuation, Bea, Runyon, and Warnock (2004) explained that local governments are the authorized units responsible for developing their locality's emergency operations and response plans. Additionally, local emergency management directors are responsible for providing training to prepare for disaster response. The importance of local planning and training cannot be underestimated in emergency management because local emergency personnel are always the first responders to disaster.

> First responders—local fire, police, and emergency medical personnel who respond to all manner of incidents such as earthquakes, storms, and floods—have the lead responsibility for carrying out emergency management efforts. (U.S. House of Representatives, 2006, p. 45)

Because local emergency personnel are on the scene first and the preparedness efforts taken by the locality, including mitigation and response training, significantly affect the ability of these emergency responders to function in the disaster environment regardless of the federal or state government's role in relief, localities bear the primary responsibility for addressing disaster preparedness efforts.

When Governor Blanco placed Louisiana in a state of emergency on the Friday before the storm made landfall, she left the authority of ordering the evacuation of New Orleans to Mayor Nagin (Shughart, 2006). According to Ripley, Tumulty, Thompson, and Carney (2005), Nagin did not issue the order to evacuate the city until Katrina was within 48 hours of making landfall, and he did not make evacuation mandatory until Sunday morning, with fewer than 24 hours remaining. The mandatory evacuation order from the Mayor came too late for immobile residents to be moved out of the vulnerable areas. Additionally, at the time the evacuation was made mandatory, most municipal workers responsible for manning pumping stations and aiding the evacuation of the immobile had already left.

> In New Orleans, the evacuation order gave little time to move residents out of the city: two hundred school buses were left to the floodwaters rather than being used to evacuate residents, municipal workers responsible for manning pump stations were evacuated rather than kept at their posts in case of a levee break. (Mulrine, 2005, as cited in Waugh, 2006, p. 13)

Ripley et al. (2005) contended that one reason the evacuation order was given so late was because the mayor was hesitant to force people out of the city; he feared being held liable for unnecessarily closing hotels and businesses. In this context, Nagin was more responsive to the possibility of political fallout in his reelection campaign in the event the evacuation was not needed than he was to the imminent danger the hurricane posed to citizens, illustrating the administration's concern for political office over the potential risk to human life.

In reference to mitigation efforts, it was discussed earlier that the city had decided to place residential development above decreasing the overall vulnerability of the population; however, the city had attempted to procure equipment needed in the event of a natural disaster but was unsuccessful. According to the Director of Homeland Security and Public Safety for the City of New Orleans, the city was limited by the federal government in their procurement requests.

> Most allowable expenditures under the UASI program remain closely linked to the WMD threat to the exclusion of many of other forms of enhanced readiness. (Ebbert, 2005, as cited in U.S. House of Representatives, 2006, p. 152-153)

Furthermore, Ebbert (2005) contends that procurement of equipment used in rescue efforts, such as flat-bottom aluminum boats, were not allowed by the Office of Domestic Preparedness because they could not be used dually in terrorist incidences. Regardless of the emphasis that the DHS places on counterterrorism activities, the General Accountability Office audited the DHS and found that out of the 36 essential capabilities first responders need to fulfill the tasks associated with responding to a terrorist attack, 30 are available in non-terrorist situations (General Accountability Office, 2005; U.S. House of Representatives, 2006, p. 153). The implications of the General Accountability Office's audit illustrated the emergency manager's inability to sufficiently interpret the Office of Domestic Preparedness' regulations regarding resource procurement, indirectly implicating that certain emergency managers are not as familiar with the office's regulations as they should have been.

Lastly, Mayor Nagin was blamed for the relief and response effort because of the manner in which he took charge of the effort. Nagin's approach to dealing with the disaster became one of inaction and displacement from the scene.

> He [Nagin] and his crisis team opted for refuge at the Hyatt Regency hotel rather than taking charge at the city's Mobile Command Center or joining other local and state officials at Louisiana's emergency operations facility in Baton Rouge. In consequence, Nagin and his advisors were cut off for two days, mostly spending their time warding off looters, as telephones went dead and the radios used by police and other first responders drained their batteries. As a matter of fact, the mayor was missing in action for most of the week after Katrina struck, busy resettling his family in Dallas. (Ripley et al., 2005, pp. 35-37)

Nagin's inaction and absence from the scene was a violation of his fiduciary responsibilities as the mayor, as discussed earlier. The lack of his presence not only illustrated his negligence in the performance of his duties, but his absence also detrimentally affected the morale of first responders. Garnett (1992) explains that a leader's presence in times of emergency aids in dispelling feelings of anxiety and, in some cases, panic.

> It is usually important for the public administrator to be on the scene, even if better facilities exist elsewhere for a command post. A leader's presence often helps calm anxieties and rally the forces coping with the crisis. A leader who sits in his or her office far removed from crisis can be viewed as uncaring, afraid, or indecisive. (Garnett, 1992, p. 208)

As a direct result of a leader's lack of presence, perceptions of the leader are sometimes created among his or her subordinates and citizens, which can potentially damage the administrator's authority and credibility (Garnett, 1992). In addition to affecting the morale of constituents, the absence of Nagin from the scene resulted in the city's emergency plan never being put into effect (ABC News, 2005; Waugh, 2006). In the midst of emergency response, communication is oriented to delegating tasks (orders, instructions, and monitoring developments); moreover, communication in reference to the "emotional health" of crisis victims is also important on the behalf of the administrator so that they are at least minimally satisfied that authorities are attempting to respond to the situation (Garnett, 1992). However, when it came to Hurricane Katrina and the inaction of Nagin, neither tasks nor victim calming rhetoric was available, resulting in an official response stalemate and a havoc-stricken and stranded population. Nagin's absence resulted in an exaggeration of the casualty situation in the city, which was used as a premise for requesting large amounts of resources.

> State and local officials, like other public sector agents, have an incentive to request a larger than efficient amount of resources when they are not the ones bearing the cost. Like other government agencies, this is generally done through an exaggeration of the true extent of the problem at hand. Both New Orleans Mayor Ray Nagin and Louisiana Governor Kathleen Blanco, for example, initially made claims that thousands or maybe tens of thousands of people were dead, with hundreds of thousands of people left trapped in their homes. In the end, however, these numbers were gross exaggerations. (Sobel & Leeson, 2006, p. 62)

Situations similar to the exaggeration of mass casualties, in addition to New Orleans' history of federal fund usage, may have contributed to the lack of action of federal agencies because agency officials did not know what to believe and what not to believe.

Effects of Katrina on the Political Landscape

The political landscape is subject to the agenda setting process. For social problems to be addressed by government departments and agencies, they first must be placed on the political agenda and then supported by a number of political actors that are capable of influencing policy changes (Anderson, 2006). Political change can occur gradually or rapidly when there is a sudden alteration to policy sentiment and issue definition (Baumgartner & Jones, 1993). According to Wood and Doan (2003), issues or social conditions can be objectively serious for long periods of time, but individuals may not publicly define the condition as a problem until the long period of serious conditions assimilate into their culture.

> An objectively serious condition may be viewed as normal by the community and evoke neutral or even favorable responses by those exposed to the condition. Values and culture produce inertia in individual perceptions and issue images. Norms and expectations develop that glue individual behavior to the beliefs of the community. As habitual acceptance of a condition permeates the community it becomes riskier for the individual to oppose the arrangement. Citizens abide by prevailing arrangements without much thought. (Wood & Doan, p. 641)

In situations such as these, social issues do not reach the political agenda and have no bearing on the political landscape because the community openly accepts the condition.

When a focusing event occurs, there are shifts in the community's social condition, which can provoke negative sentiments from the people in reference to their past social conditions. Kingdon (1995) explained that although policy windows can occur due to institutionalized events, such as elections and budgetary cycles, they can also occur because of the influence of focusing events, crises, and accidents. In the event of a focusing event, a policy window opens because social conditions capture the attention of government officials and citizens close to the conditions (Kingdon, 1995). Under circumstances where focusing events play a major role in issues being placed on the political agenda, the government tends to play an active role in defining problems and setting goals regarding how to address the problems (Jones, 1984; Fiorino, 1990).[24] When focusing events are caused by natural disasters, whose occurrences are relatively unpredictable, the policy window that opens is referred to as a *random policy window* (Kingdon, 1995; Howlett, 1998).[25] In reference to New Orleans, Katrina created a random policy window through which local and federal policies could be changed; however, it is still currently unclear what progress has been made in addressing the social conditions discussed in the previous chapter.

The effect of a catastrophic disaster on the political landscape is usually directly related to officials' actions or inactions and the official's overall effectiveness (Abney & Hill, 1966). Disasters place new issues on political agendas that usually reference "mistakes" made by incumbents, who must alter their reelection campaigns to compensate for their errors in responding to the hazard event (Twigg, 2004). Disasters bring previously unaddressed or low salience issues to the center of political discussion (Carmines & Stimson, 1993), making the disaster a focusing event where policy changes can occur more easily than they could prior to the event's occurrence (Kingdon, 1995). In the case of New Orleans, Katrina added salience to the issues of racial marginalization; however, movement of this issue is questionable. Although a disaster may create a policy window, what makes a disaster a focusing event is the reaction to the event on the behalf of constituents. If the marginalized residents of New Orleans do not engage in political activism or protest in reference to their marginalization, there will be no progress; however, political expression even under these premises may have a limited effect because of the possible tone of political expression.[26] In the aftermath of a disaster,

> Symptoms of aggression become highly visible among a population that is recovering from a disaster. . . . The readiness to give vent to angry resentment and heated condemnation, particularly of local officials, is perhaps the most widely noted manifestation. (Janis, 1945, as cited in Abney & Hill, 1966, p. 974)

Additionally, Form et al. (1956, as cited in Abney & Hill, 1966) explained that resentment on the behalf of a population is usually concentrated in the lower classes and is manifested in hostile rhetoric or actions toward middle and upper class segments of the population, institutions, and organizations that operate after the disaster when they cannot.[27]

Although disasters have the ability to act as focusing events and create policy windows, disaster problems recede from the political spotlight with the passage of time, and agendas revert back to what they were before the event, diminishing the chance for further change to occur (Baumgartner & Jones, 1993; Jones, 1984; Twigg, 2004). In addition to time relevance, the type of natural disaster effects whether there will be changes to the political landscape. According to Birkland (1996, 1997), hurricanes have a more limited effect on national agendas than other events because more knowledge exists about hurricanes than earthquakes or nuclear disasters. Hurricanes occur more frequently than other disasters, which results in less media exposure and legislative alarm (Twigg, 2004). Although hurricanes tend not to have significant effects on the national political landscape, hurricanes have more significant effects on the political landscape where they strike (Twigg, 2006).

Before a natural disaster, elected officials and citizens share the perception that officials' day-to-day role is custodial. During a disaster response, relief, and the reconstruction process, citizens believe that officials should be more active even though officials continue to retain their custodial perception (Twigg, 2004). When perceptions of responsibilities on behalf of citizens and officials do not match, there can be potential electoral effects. According to Twigg's *negative effects theory*, a major hurricane or other disaster event can potentially focus blame on the elected official, encourage strong opposition, and potentially result in electoral defeat; however, this does not necessarily have to occur. Abney and Hill (1966) argue that the political repercussions of a natural disaster are not automatically detrimental to the government in power.

For example, when Hurricane Betsy struck Southeastern Louisiana in 1965, and created flooding, destruction, and death, the mayor of New Orleans, Victor Schiro, was reelected even though he was politically attacked for inadequate preparation and response to the hurricane (*The New Orleans Times-Picayune*, 1965, as cited in Abney & Hill, 1966). The reelection of Mayor Nagin after the passage of Hurricane Katrina also illustrates this point; however, the outcome of the election does not necessarily reflect the political sentiments of the entire New Orleans population prior to the storm.

The election's reflection of local political sentiment is questioned by most residents because they were dispersed across the country at the time of the elections. Furthermore, those living in the "dry" universe had no reason to think that the response and relief that took place in the aftermath of Katrina was detrimental. Therefore, in the absence of voters disillusioned by Nagin's behavior during the relief and response process, he was voted back into office by those still living in the area who believed he did a decent job because they were relatively unaffected by the flooding anyway.

In addition to potentially changing government administrations, disasters create policy windows for altering local social services and conditions (Twigg, 2004). In the case of New Orleans, as discussed in Chapter 3, the social conditions and services available to the majority of people living in the city prior to the storm were already in dire need of improvement. The future resilience of New Orleans depends on repairing the social infrastructure of the city just as much as it does on repairing the physical infrastructure. Political action and agenda setting took place the same day that the levees failed with lobbyists in Washington, D.C., soliciting the federal government for aid in the rebuilding process (Congleton, 2006). In response to the heavy lobbying efforts after Katrina, within a month, seven emergency relief bills were passed by both chambers of Congress and signed by the President. Included in these were two bills, appropriating a total of $60.23 billion for new spending, three bills appropriated

Ineffective Leadership

Figure 4.1. Reprinted with permission from Steve Kelly, *The Times-Picayune*.

funds for aiding "needy families and students," a $2 billion increase to the borrowing limits of the National Flood Insurance Program, and a tax relief bill that reduced taxes by $6 billion in the 2006 to 2010 period (Congleton, 2006). In addition to fund appropriations at the federal level, in an effort to change the social conditions of the city and instill a sense of empowerment in its residents, the notion of a more participatory government has emerged.

> Although this emphasis is driven by a sense of justice, government also functions most effectively when it works in partnership with community groups that can provide local knowledge, mobilize resources, recruit volunteers, and highlight urgent issues that easily fall below technocratic and regulatory radar. (Pastor et al., 2006, p. 35)

The pursuit of structures that would make it possible to integrate residents into the decision-making processes of New Orleans' city government is commendable; however, past attempts in this vein have been procedural in practice and have not ensured equitable outcomes in regulatory, zoning, land use planning, economic development, or facility site decisions.

Although more resident participation in government decision making may empower those more vulnerable and detrimentally affected by the storm, if the

The Budget Motel

Figure 4.2. Reprinted with permission from Steve Kelly, *The Times-Picayune.*

city is reconstructed in a fashion that does not include traditionally marginalized groups, participation will remain as it was prior to the storm. According to Dynes (1970), people remaining in the area affected by disasters tend to be frequently optimistic about the rebuilding process and the future of their neighborhoods and city. Recommendations for rebuilding range from rebuilding communities similar to how they were prior to the storm to not rebuilding large parts of New Orleans at all to moving most of the development away from the coastline of the Gulf Coast (Dean, 2005).

The absence of rebuilding sections of the city seems to be a benevolent aspect that directly relates to avoiding future vulnerability to disasters; however, this will discourage segments of the population who lived in those neighborhoods from moving back into the city. Additionally, reconstruction necessitates the temporary housing of workers, as discussed in the third chapter, is composed mainly of illegal Mexican immigrants; this will alter the demographic composition of the city in the future and will affect voting patterns. However, if FEMA trailers that have been promised for temporary housing are not totally distributed to whomever is in need (prior residents or new workers), there will be little for anyone to move back to, at least in the immediate future.

Summary

The responses to Katrina in the political landscape have the potential to reshape the topography (i.e., new policies) by reconstructing more earthen and manmade levees and flood walls. Just as specific places create the potential for certain experiences, places also create the potential for alterations of the political landscape. In reference to New Orleans, the location dissuaded human development into certain areas; however, the political landscape that developed actively sought to overcome natural limitations posed by annual flooding. The political landscape continually places segments of the population at risk from natural hazards in the interests of wealth and personal political recognition.

As a focusing event, disasters create a policy window where social concerns can be addressed. Hurricane Katrina exposed social issues of segregation, poverty, and minority marginalization. Moreover, the storm highlighted the problems that plague the bureaucracy, the presidency, and state and local government officials with regards to disaster mitigation, preparedness, and relief. It was expected, in theory, that due to the ineffective response and relief, the political landscape would have been extremely altered by both government policies made in reaction to the event and new voting patterns; however, this was not the case. Overall, the political landscape of New Orleans has changed little.

Although the notion of hazard preparedness is on the minds of officials, it will soon fade away like it has so many other times. The social concerns of poverty, segregation, and other aspects of minority marginalization are unlikely to be addressed by officials because the people who were detrimentally affected by these circumstances are no longer present within the political landscape, thereby taking away incentives for officials to address the issues. Although Katrina caused so much destruction, it was hoped on the behalf of marginalized residents that there would be some indirect benefit from the disaster in the form of political attention to their social plight; however, it is possible that such a change to the political landscape will not come, or rather, it will not be influenced by dramatic natural events.

Notes

1. Levees were preferable to alternative land use policies because the construction of levees allowed for more available land that could be developed by local governments.

2. "The Mississippi Flood of 1927 lasted for two months and covered an area of 27,000 square miles" (Rivera & Miller, 2006, p. 7).

3. See also McQuaid and Schleifstein, 2006

4. See also Cutter, 1996; Cutter, Boruff, and Shirley, 2003; and Cutter, Mitchell, and Scott, 2000.

5. Cutter and Emrich (2006) asserted that if a community is stressed economically and losing population prior to a disaster, the trend will continue long after disaster relief and reconstruction are complete.

6. The other primary component of place vulnerability deals with the characteristics of the people and the places that make them less able to cope with and rebound from disaster events, such as socio-economic characteristics and urbanization (Cutter & Emrich, 2006).

7. "Subrogation occurs when an insurance company which pays its insured client for injuries and losses then sues the party which the injured person contends caused the damages to him/her" (Law.com, 2007; see also Burby, 2006).

8. "Moral hazard is an insurance term that refers to cases where the availability of insurance protection lowers an insured party's incentive to avoid risk. Insurance companies try to counter this through the use of deductibles, higher insurance rates, and the threat of canceling policies if claims are too frequent" (Burby, 2006, p. 180).

9. The enactment of the Disaster Mitigation Act of 2000, in conjunction with the National Flood Insurance Reform Acts of 1994 and 2004 and federal aid that reduced the cost to local governments in reference to providing infrastructure in hazardous locations, has stimulated residential and industrial development in these same locations (Burby, 2006).

10. According to Shughart (2006), each of these boards were composed of both gubernatorial and local political appointees who had broad taxing and borrowing powers in order to support financial obligations incurred through their duties.

11. "A National Science Foundation-funded team sent to investigate New Orleans' post-Katrina flooding concluded that many of the weak spots breached by the storm resulted from unclear lines of authority and insufficient coordination amongst the various agencies having jurisdiction over the levee system. Flood-walls were built of different heights in some locations and of different, ineffectively joined materials in others. At one pumping station, for which at least three agencies potentially were responsible, for example, a concrete floodwall connected to an earthen levee that was much lower. Katrina's storm surge over-topped the shorter structure, rendering the more substantial one useless." (Carrns, 2005, as cited in Shughart, 2006, p. 35)

12. Some blame scenarios include private and nonprofit organizations.

13. Gormley and Balla (2004) used the EPA as an example of how intergovernmental agencies, similar to FEMA, had difficulty in implementing a policy that reported on itself nationally: "The EPA, for example, depends on states to implement numerous programs aimed at reducing water pollution. Because states measure water quality in many different ways, it is difficult for the agency to know how much progress has been made over time and across different areas of the country" (p. 17).

14. Out of the 18 agencies that were strung together to create the Department of Homeland Security, FEMA was the only independent agency.

15. Catastrophic incidents are "any natural or manmade incident, including terrorism, which results in extraordinary levels of mass casualties, damage or disruption severely affecting the population, infrastructure, environment, economy, and national morale and/or government functions. A catastrophic event could result in sustained national impacts over a prolonged period of time; almost immediately exceeds resources normally available to State, local, tribal, and private sector authorities; and significantly interrupts governmental operations and emergency services to such an extent that national security could be threatened" (U.S. Department of Homeland Security, 2004, as cited in Sylves, 2006, p. 28).

16. See also Waugh, 2005, and Waugh and Sylves, 2002.

17. The National Response Plan "organizes capabilities, staffing, and equipment resources in terms of functions that are most likely to be needed during emergencies, such as communications or urban search and rescue, and spells out common processes and administrative requirements for executing the plan" (U.S. House of Representatives, 2006, p. 32).

18. The National Incident Management System "consists of six major components of systems approach to domestic incident management: command and management, preparedness, resource management, communications and information management, supporting technologies, and ongoing management and maintenance. According to the DHS,

NIMS 'aligns the patchwork of federal special-purpose incident management and emergency response plans into an effective and efficient structure'" (U.S. Department of Homeland Security, 2004, as cited in U.S. House of Representatives, 2006, p. 32).

19. See also Kettl, 2005, and Cigler, 2005.

20. Examples include actions taken by the U.S. Coast Guard, a Canadian search and rescue team, and a county sheriff from Michigan (see Phillips, 2005, and Parker, 2005).

21. "FEMA's top three leaders—Director Michael D. Brown, Chief of Staff Patrick J. Rhode, and Deputy Chief of Staff Brooks D. Altshuler—[had] ties to President Bush's 2000 campaign or to the White House advance operation, according to the agency [FEMA]. Two other senior operational jobs are filled by a former Republican lieutenant governor from Nebraska and a U.S. Chamber of Commerce official who was once a political operative" (MSNBC, 2005, as cited in Sylves, 2006).

22. "By way of contrast, when Hurricane Charley struck the electoral-vote-rich, battleground state of Florida the previous August—and with his own reelection campaign in full swing—the President was on the ground two days later (CNN, 2004). . . . President Bush seized the photo-op moment in 2004, but waited a full four days before visiting Katrina's impact area" (Shughart, 2006, p. 38).

23. The official evacuation phase occurred between 4:00 p.m. on Saturday to 6:00 p.m. Sunday (Audio Recordings of Hurricane Katrina Conference Calls, 2005a, as cited in U.S. House of Representatives, 2006).

24. Fiorino (1990) lists the effects of Love Canal, Bhopal and Exxon Valdez as events that generated public concern that provoked proactive action on the behalf of government entities.

25. Howlett (1995) describes Kingdon's (1993) four principal policy windows to include routine political windows, discretionary political windows, spillover problem windows, and random problem windows.

26. According to Miller, Rivera, and Yelin (forthcoming), the possibility of political protest on the behalf of a disgruntled segment of the population may be legally cut off before making any progress in building salience for the issue because of the government's new fear of government criticism.

27. Abney and Hill (1966) further differentiate between segments of the population more inclined to be hostile to elected officials in the aftermath of poorly managed disasters in New Orleans by explaining that residents live in two different universes: the wet and dry. "The 'wet' section where people had the traumatic experience of being flooded out of their homes" (p. 975) and the "dry" section where residents were not flooded out and carried on their existence relatively normally after the disaster.

Utter Destruction

Obliteration

A view of the Lower 9th Ward. A visitor to the Lower 9th Ward views the devastation caused by the storm surge and the flooding after the levee ruptured.

Hazards in the disaster landscape. The disaster landscape is free of risks. It exposes those who return to rebuild to a new environment in disarray.

Now Hiring with Bonus

Disturbed cemetery sites. Cemeteries were disturbed throughout the region. As can be seen in this photograph, the casket has come out of its crypt.

The disturbed dead. Tombs moved by the force of high waters in the Jackson-Square-Harrison Cemetery located in Slidell, Louisiana.

Steps Leading Nowhere

Indiscriminant Destruction

Where a House Once Stood

Chapter 5
Views of Changing Landscapes

When Hurricane Katrina struck New Orleans, dozens of people stayed alive by setting up camp among the graves of a cemetery, where they lived for days with no sign that help would come.

Nellie Francis, 77, was one of the residents of the makeshift camp at Mt. Olivet Cemetery, where mausoleums served as shelters and people set up their own emergency government, running rescue efforts, tending to the sick, and feeding the hungry—in short, filling the void left by a lack of noticeable response to the disaster.

—Christy Oglesby, 2005

Hurricane Katrina offers lessons not only in environmental change but also in social change. Not only were there several physical changes, but Katrina has also set the stage for changes in both the processes and social structures that have yet to unfold. "Disasters unmask the nature of a society's social structure, including the ties and resilience of kinship and other alliances; they instigate unity and the cohesion of social unity as well as conflict along the lines of segmentary opposition, [and they ultimately help us understand] the distribution of power within a society" (Oliver-Smith & Hoffman, 2002, p. 10). Because disasters are unique in their onset and the cultural understanding of the events varies, no one can truly predict the course of changes that will take place after the hurricane has passed. This chapter concerns itself with understanding the scale of devastation and analyzes human trauma in terms of the toll exacted when social and political systems fail to meet the needs of survivors.

Furthermore, this chapter presents views of how classical and contemporary social thinkers conceive societal change on a variety of individual, organizational, and societal levels. Because "sociological research naturally help[s] us to identify and understand what goes on when disaster strikes but also, conversely, the investigation of these phenomena can extend sociological theories of human behavior and social organization" (Merton, 1969, p. xi), we are better equipped to understand the current disaster and the impact ecological devastation has on society as a whole. Also, this chapter stresses the importance of the symbolic connection with the environment and how it shapes social change in communities in the aftermath of disaster.

Community Redevelopment

As survivors get caught in the milieu of the socio-economic consequences of disaster, they often get lost in the discussion of the post-disaster recovery efforts. Recovery and change take place on a multitude of levels and vary greatly in scale with each disaster. For instance, in the work *The Resilient City: How Modern Cities Recover From Disaster* by Vale and Campanella (2005), the authors argue the nature of cities that have survived disaster and gone on to recover from traumatic experiences. *Comeback Cities* by Grogan and Proscio (2000) details how neighborhoods, main streets, community organizations, and local leaders can combine forces to implement changes that can affect the entire city one institution at a time. Fried's (1966) classic essay, *Grieving for a Lost Home*, examines the losses suffered by individuals displaced from Boston's West End Community, which expresses the effects disasters have on a personal level. These three works range in type of disaster or trauma that occurred as well as in terms of ecological assault and their level of impact.

The work of Vale and Campanella (2005) focuses on micro-level adjustment to disaster as well as social trauma and change; the work of Grogan and Proscio (2000) focuses on inner-city social change that ultimately affected the city-wide socio-economic structures; and the work of Fried (1966) focuses on micro-adaptive changes needed to cope with the loss of a home, which is often a symbol of security and safety and is tied to one's identity. With change occurring simultaneously on several levels of society, it is important to understand how these changes affect one's sense of place and the overall recovery process.

Place Attachment, Communities in Transition, and Social Change

The place attachments that develop over time can all be washed away with a flood or blown away in a hurricane or tornado. Natural disasters disrupt daily routines with the sounds of official warnings, knowledge of evacuations, images of neighbors suffering or dying, and extended shelter stays that can break down social institutions if not managed correctly. So traumatic are these events (as later accounts from survivors of Hurricane Katrina will detail) that many argue that the long-standing bonds to a place are forever changed. According to Brown and Perkins (1995), "Place attachments develop slowly but can be disrupted quickly and can create a long-term phase of dealing with loss and repairing or re-creating attachments to people and places. These . . . phases are interdependent as qualities of the initial attachment or disruption can ease or escalate the stress of loss and the difficulty of re-creating attachments" (pp. 284-285). The problem with understanding changes in the layered landscape is reconciling the psychological, social, and physical aspects of recovery amid a physical landscape in disrepair. In fact, certain aspects of the pre-Katrina landscape may no longer exist (i.e., coastal erosion along Lake Pontchartrain) to give survivors the opportunity to effectively recover.

With abrupt disruption of the landscape, a multitude of reactions can occur, including shock, dismay, psychological depression, anger, disability, and fear, and can linger for months and sometimes years. These all contribute to what researchers call the *Katrina malaise*. Much of the Katrina malaise is the result of a series of failures the communities experienced after Katrina (Lotke & Borosage, 2006; Picou & Marshall, 2007). These failures include a failure to prepare, a failure to respond, and a failure to rebuild. Ultimately, disasters reveal the differentials among group resiliency and the allocation of resources for reconstruction.

The nature of the immediate disruptions coupled with the involuntary relocations violated the residents' assumptions that the levee system would keep them safe. As survivors of Hurricanes Betsy[1] and Camille, residents were convinced they would be safe. Levees and homes that were once understood to be safe havens "became weapons [during Hurricane Katrina], trapping neighborhoods in the flood and dismembering others" (Brown & Perkins, 1995, p. 291). So forceful were the waters that flooded the Lower 9th Ward that many homes could not withstand the sheer force and were completely destroyed or were transported by the waters several yards away from their foundations, some even stopping atop cars. Survivors quickly realized that the technological failure at the levees unleashed a series of changes. Several months after Katrina and the recession of the waters, search and recovery missions retrieved hundreds of bodies from the attics of homes in New Orleans. The society and culture of New Orleans may never be the same again.

As if the event itself did not cause enough disruption, the post-Katrina disruption—the government bureaucracy, leadership ineptness, and (what some consider) utter contempt for those suffering in New Orleans—made matters worse. The lack of an immediate, coordinated recovery called into question the "old order" and the government's ability or willingness to learn from earlier natural disasters of national importance.

Social Disruption and Views of Changing Landscapes

For centuries, urban areas have endured calamities such as fires, hurricanes, floods, and earthquakes. Conventional wisdom assumes that lessons from past disasters would be put into practice. Moreover, this conventional wisdom, with a background stemming from at least seven major disasters[2] and numerous minor ones, assumes that a culture of preparedness would develop to address the fundamental issues of disaster mitigation. A culture of preparedness would go beyond the preparation of disaster plans and move toward a culture that is ready to handle disasters as they come. Based on prior disasters, such a culture would include a respect for different opinions and groups in the redevelopment phase of seeking support from various stakeholders. Civic stakeholders argue that such changes must be made so survivors can envision a future.

Amid changed social, political, economic, and cultural landscapes, what will become of New Orleans in the aftermath of Katrina? "Will it be an embittered ghost town like the towns in Georgia that never recovered from Sherman's March? Or will it rise from the ruins like Atlanta—and become a world plan part two?" (Hutchinson, 2005). A member of Mayor Nagin's administration is quoted in the online version of the *Wall Street Journal,* where he speaks of the need for a "new" New Orleans:

> The power elite of New Orleans ... insist the remade city won't simply restore the old order. New Orleans before the flood was burdened by a teeming underclass, substandard schools and a high crime rate.... The new city must be very different, Mr. Reiss [a member of Mayor Nagin's administration] says, with better services and fewer poor people. "Those who want to see this city rebuilt want to see it done in a completely different way: demographically, geographically, and politically," he says. "I'm not just speaking for myself here. The way we've been living is not going to happen again, or we're out." (Cooper, 2005)

The above quote echoes the sentiment for change and issues a call for a development that lies at the root of the opportunity citizens have to restructure life after disaster. As noted, it is a call to completely restructure the city by way of geographic, political, and demographic changes that will result in a more just and equitable city for all its inhabitants.

> Understanding the physical environment and the symbolic connection with the environment individuals have and how it shapes exchange relationships, redevelopment processes, cultural expression are all the key in the study of extreme events and social change as communities experience extreme trauma. More precisely, when technological failures disrupt the relationship between the community and environment, both will experience change. These social changes unfolding in "extreme environments provide significant opportunities for sociological analysis ... since the ensuing conditions of individual and collective stress expose usually concealed institutional behavior to observation and examination and reveal and magnify aspects of social systems and processes that are typically obscured by the routinization of everyday life." (Kroll-Smith, Couch, & Marshall, 1997, p. 4)

The survivors' everyday routines are changed just as the landscape is changed. Constant exposure to risks induces socio-cultural disruptions to once stable institutions that form the foundation of civic life.

Historically, researchers have categorized traumatic events and charted the process of change society undergoes as a result of precipitous events such as Hurricane Katrina. Carr (1932) noted four types of changes in the following pattern: (1) population changes, (2) cultural changes, (3) relational changes, and (4) catastrophic changes.[3] Also, disasters can be conceptualized and differentiated in terms of their speed, scope, complexity, and the changes in social behavior that occur. For example, social scientists have extensively documented the community changes in the aftermath of a natural disaster. Following a natural disaster, communities are typically characterized by a spirit of cooperation in which individual members tend to "pull together" and work to address the issues at hand. However, after a technological disaster or a disaster that has both natural and technological aspects, researchers have witnessed the emergence of a phenomenon called the corrosive community, which is characterized by social disruption, a lack of consensus about the environmental degradation, and a general uncertainty about the future (Cuthbertson & Nigg, 1987; Freudenburg &

Jones, 1991; Gill, 1994; Gill & Picou, 1998; Kroll-Smith, 1995; Kroll-Smith & Couch, 1991, 1993a, 1993b; Ritchie, 2004).

More specifically, social corrosion is a "consistent pattern of chronic impacts to individuals and communities" (Picou, Marshall, & Gill, 2004, p. 1496). Furthermore, this long-term pathological trend has been evidenced by the breakdown of social relationships, the fragmentation of community groups, family conflict, loss of trust, prolonged litigation, and the use of self-isolation as a primary coping strategy (Arata, Picou, Johnson, & McNally, 2000; Picou et al., 2004; Picou & Marshall, 2007). Picou and Marshall (2007) also contend that "there is a lack of sympathetic and empathetic behavior for survivors from non-victims, and both resources and support capabilities from local, state, and federal institutions decline over time, and these processes result in the continued deterioration of the social organization and culture of impacted communities, and residents experience group and interpersonal conflict as well as severe psychological problems" (p. 13).

To understand social change within the layered landscape framework, three perspectives, in their broadest sense, are considered. These perspectives derive from different historical traditions, world views, and epistemological backgrounds; they offer insights about the three domain assumptions that arise, namely the functions or dysfunctions that occur in a society within each landscape and how change occurring in one landscape influences change in the others. The constant state of conflict and tension in a society, the interpretation of processes, and the social construction of change offer insight when explaining the adaptation processes after a disaster. According to Harper (1998), these basic assumptions lead to the following question: "What factors determine the structure of society and the nature of their change?" (p. 88). Although different perspectives are presented, the current work employs the assumption that landscapes are layered spaces, and the interpretive approach, namely the ecological symbolic approach, is used to better conceptualize the changes to one's connection to place attachment occurs.

The Functionalist Approach

The functionalist approach argues that society and social change are brought on by the necessities of survival. During times of crisis, the local residents who made the makeshift camp in the Mt. Olivet Cemetery bonded together to facilitate the necessities of survival. The Mt. Olivet Cemetery group was able to establish a temporary community to compensate for absent federal authorities, who were overwhelmed by the magnitude of the crisis and unaware of the individual problems. With environmental and organizational changes occurring at such a rapid pace and the problems with governmental agencies running so rampant that a dysfunctional crisis emerged (a situation that many functionalist theorists would argue occurred under these circumstances), the survivors were isolated, and they were unable to remain integrated within the larger social structure around them. Hence, a sub-unit developed to relieve the temporary needs of the Mt. Olivet Cemetery group. This replacement, albeit a short-term set of roles needed for the survival of the group, is what Parsons[4] (1966) referred to as *adaptive upgrading*.

In essence, in isolation, their social system became more effective in generating and distributing resources, thus enhancing the group's survival. However, this temporary change did not last and was soon replaced by the traditional social structure that was re-established with the arrival of search and rescue teams. Although the Mt. Olivet Cemetery group survived by altering their way of life to retain some sense of normalcy, they did not fundamentally alter the basic values, goals, distribution of power, internal patterns of order, or any other aspect of the American disaster relief efforts. For lasting change to permeate the society there must be an overall alteration of abstract core values, both in terms of citizen evacuation patterns and the government's responsibility in meeting the needs of citizens in a time of crisis.

More recent functionalist views of social change define society not so much in the classical sense, where society is seeking equilibrium, but rather as a tension-management system (Moore, 1974). Social change is described by Olsen (1978) as a constant move toward adjustment:

> Whenever stresses or strains seriously threaten the key features of an organization—whatever they might be—the organization will . . . initiate compensatory actions to counter these disruptions, in an attempt to preserve its key features. If the compensatory activities successfully defend the threatened key features, then whatever changes do occur will be confined to other, less crucial features . . . to the extent that the organization successfully practices such adjustive maneuvers, it survives through time a relatively stable social entity. . . . When disruptive stresses and strains or their resulting conflicts are so severe and prolonged that compensatory mechanisms cannot cope with them, the key organizational features being protected will themselves be altered or destroyed. The entire organization then changes; there is a change of the organization rather than just within the organization. (p. 34)

The functionalists explain how social change takes place, but their theories of change tend to be more gradual in nature. One major criticism in the gradual approach is that society progresses into a state of equilibrium. However, in the wake of destruction that alters the physical, social, economic, and political landscapes, there is rarely any time for the gradual progression into a homeostatic society. Although this argument for social change explains some aspects of change, functionalism and neo-functionalism alone cannot adequately address the abrupt nature of the multitude of coping strategies for the recovery process to move forward in the midst of radical, abrupt landscape changes.

The Conflict Approach

A second group of theories purport that social change is an outgrowth of inescapable competition for scarce resources in a society. Here, conflict is considered inevitable in social systems; it is seen by some as a creative source of change and by others as the only means of change. In fact, for Karl Marx, conflict is a normal part of life, and conflict and social change are inseparable. According to Marx, the events that lead to change in a class struggle during a proletarian revolution are as follows: "(1) the need for production, (2) the expansion of the division of labor [in a society], (3) the development of private property, (4) increasing social inequality, (5) class struggle, (6) the creation of political structure to represent each class's interest, and finally, (7) revolution. Each event leads inevitably to the next event" (Duke, 1976, p. 28).

Prior to Katrina, the city's economic, cultural, and social landscapes were filled with struggles caused by a number of disparities[5] ranging from healthcare inequalities to employment and hiring inequalities. These inequalities are rooted in a long history of racial and ethnic disparities but relate to the saving of lives in the wake of a natural disaster. In reference to marginalized groups, the events that occurred after Hurricane Katrina were not isolated events in United States' history; as Pastor et al. (2006) contend, "Katrina is only part of a long run historic record of inequality in disaster vulnerability" (p. 22). For example, in 1822, hundreds of slaves died when a hurricane made landfall in South Carolina because there was no high ground and no shelter for the slaves to use (Mulcahy, 2005). Moreover, the 1927, Mississippi Flood took the lives of hundreds of Blacks who were rounded up and put on levees without food, water, or shelter. White authorities did not allow them to evacuate because they feared they would lose their inexpensive labor force (Barry, 1997).

In 1928, a major hurricane hit South Florida and more than 2,500 people, mostly Black migrant workers, drowned in what is considered one of the worst disasters in U.S. history (Gross, 1995; U.S. Weather Service, 2006; Van Orden, 2002). During Hurricane Audrey, which made landfall in Louisiana in 1957, the death rate was significantly greater for Blacks (Bates, Fogleman, Parenton, Pittman, & Tracy, 1963). Research conducted in the 1970s concluded that disaster-connected deaths were disproportionately high among ethnic minorities (Trainer & Hutton, 1972), and research on loss from natural hazards in the United States from 1970 to 1980 further confirmed that lower income households experience higher rates of injuries in disasters such as floods, earthquakes, and fires than more affluent households (Rossi, Wright, Weber-Burdin, & Pereira, 1983). The pattern of differential impacts is often due to the quality of housing afforded those lower on the socio-economic scale.

The poor quality construction of low-cost housing puts residents at greater environmental risk (Aptekar, 1990; Bolin, 1986; Bolin & Bolton, 1986; Greene, 1992; Phillips, 1993). This well-documented, pervasive inequality has been long noted by scholars from a diverse group of academic fields (including sociology, economics, anthropology, and geography) who recognize that race, ethnicity resources, income, gender, ability, status, and age all shape the readiness for disaster, the response to disaster by the public and by civil authorities, and the consequences of disaster. "When the hurricane hit, the existing inequalities and the history of discrimination in the American South played out in tragic yet predictable ways" (Pastor et al., 2006, p. 3). For example, evacuation strategies left the most vulnerable populations—the poor, minorities, and the elderly—inadequately protected. Bruce Nolan, a *Times-Picayune* reporter, summed up the emergency transportation plan eloquently: "City, state, and federal emergency officials are preparing to give the poorest of New Orleans' poor a historically blunt message: In the event of a major hurricane, you're on your own" (2005).

Social Change and Lefebvre's Ideas of Space

Lefebvre's theory[6] of space is the idea of an indefinite number of spaces piled one on top of one another that can be analyzed by different disciplines. The physical (natural), mental (logical and formal abstractions), and social are the three fields Lefebvre contends we need to understand by uncovering their theoretical unity. Lefebvre organizes an understanding of spatiality around separate concepts of space: absolute space is the natural environment until it is colonized, at which point it becomes relative and historical; abstract space is the location wherein production and reproduction processes are separated and space takes on an instrumental function; contradictory space is where the breakdown of the old and the build up of the new occur; and differential space is the consequent mosaic of different places (as cited in Dear, 1997). Also, within this conception is the idea that social production is the act of producing space (i.e., "[social] space is a [social] product" [Lefebvre, 1991, p. 26]). For Lefebvre (1991), the following four precepts are pivotal:

1) Physical space is disappearing.

2) "Every society, every mode of production, produces its own space" (p. 31) wherein all social space contains and assigns appropriate places to the relations of production and reproduction of labor power and social relations as well as biological reproduction. The problem with this is that certain spaces became domestic spaces and are able to mold subordinate spaces on the edges or periphery (Dear, 1997).

3) "Theory reproduces the generative places" (Lefebvre, p. 37). In essence, if space is a product, the knowledge of it will reproduce and expand the process of production. Lefebvre emphasized the dialectic nature of identity by distinguishing among our spatial practices (our conceptional) and representational space (the lived space).

4) The passage from one mode of production to another is of the greatest importance (Lefebvre, p. 46). When one mode of production in society shifts to another, it is necessary for the new production to have its own space.

According to Dear (1991), Lefebvre assumes that space is present and implicit in the act of creativity and being and that the process of life is inextricably linked with the production of different spaces with many sociopolitical implications. In essence, those who control and dominate the production of space also control how the space is used. Thus, the conquest for space is a part of the overall mode of production that changes the complex nature of the interaction patterns within landscapes. Social change occurs radically with the making and remaking of space and the formation of place attachment as the use of a place is discontinued. Those in control change through the practice Lefebvre called *diversion* (detournement), the rise of the physical landscape, and often through the physical and economic landscapes. Natural disasters afford the perfect opportunity to divert (detournement) the function and purpose of the physical landscape and redefine its use for explorative purposes. However, it is important to understand how people interpret their lived experiences within their layered landscapes.

The Interpretive Approach

The interpretive theories view the institution as a complex system that undergoes a series of modifications depending on how different groups define their circumstances. Change is the result of the understanding that the plurality of definitions has a shared common meaning. Change becomes evident after contested meanings become problematic for others in the social organization. In other words, when the meanings that have been assigned to social institutions no longer work, humans may discard, modify, or create symbols to represent their world in a more palatable way. From an interpretive perspective, meaningful social change happens when individuals redefine situations and then act on those meanings. However, interpretive theories of social change are less deterministic in nature and do not tell us much about any "structural" sources of such redefinitions.

One such interpretive theory useful in understanding and explaining change as a result of environmental catastrophe and landscapes stricken by disaster is the ecological-symbolic perspective. According to Kroll-Smith et al. (1997), the ecological-symbolic perspective theorizes about the material and non-material properties of the physical world and has two basic propositions: first, "people live in exchange relationships with their built, modified and biophysical landscapes/environments," and second, "disruptions in the ordered relationships between communities and landscapes/environments are locally interpreted and responded to as hazards and disasters" (pp. 6-7).

In the first statement, Kroll-Smith & Couch (1991) maintain that the interdependent reality of the interaction between humans and the physical world encourages researchers to view disaster impact and response by looking at how types of aversive environmental events alter the relationships between people and their environments. The second axiom assumes that both the nature of the disruption in human-environmental relations and the appraisals people make of those disruptions affect recovery. In short, disasters are subjectively comprehended changes in the physical destruction and the symbolic capacities of humans. It is not that people derive meaning directly from nature but rather that the radical changes in nature initiate a need to interpret or understand it. Although there are many problems with this dialectical approach, including the approach's lack of measures that can predict social change, it does make it possible to "conceive of systems encompassing both human and physical elements" (Duncan, 1961, p. 141).

The ecological-symbolic approach holds that responses to an event such as Hurricane Katrina are shaped by both the nature of the environmental disruption and how the disruption is apprehended *via* interpretative frames. "The interpretive processes mediates how humans experience environmental trauma and that these process[es] are influenced by the type of environment that is damaged"

(Kroll-Smith & Couch, 1991, 1993a, 1995, as cited in Ritchie & Gill, 2007, p. 112). Communities exist as a series of interconnected relationships built on norms of reciprocity not only with other humans but also with the environment. It is important that when an exchange relationship between humans and the built and natural environments is physically devastated, a renegotiated consensus emerges as a result of natural or technological disasters. Such stressors can negatively affect the community's relationship with the environment.

These disruptions between humans and their environments cause trauma. Trauma resulting from technological disasters create collective stress, including cultural change, which involves "reality disjuncture" (i.e., no shared group assumptions), and structural change, which disrupts a community's routines and social networks (Kroll-Smith & Couch, 1993b). "These sources of stress generate additional stressors because of accompanying uncertainty, loss of control, alienation, and issues surrounding threat belief systems" (Ritchie & Gill, 2007, p. 112). For many survivors who entered the city in the immediate aftermath of the devastation, the landscape appeared foreign to them.

A Refuge in Gentilly

At St. Gabriel the Archangel Church in the Gentilly Woods area of New Orleans, worshippers gather in a church that has become a refuge for Aria Bocage, a 26-year-old mother of two. She and her husband are picking up the pieces in this devastated neighborhood. "You step into the doors of the church, it's like nothing ever happened. But soon as you step outside, you look across the street, you see the homes are still battered."[7]

> Each Sunday, Mary Gold Hardesty notices "8 or 10 people" newly returned to the neighborhood. For her, the resurrection of St. Gabriel from the flood is a sign that her middle-class, largely African American neighborhood will be reborn, too. In fact, all but two of the homeowners on her *cul de sac* of Mendez Street are back. (Ydstie, 2006)

It is the ability to see the landscape in total disarray and socially reconstruct reality as a result of cognitive frames and the interpretation of those frames that signals the rebirth of the Gentilly neighborhood. The church represents hope amid a landscape in turmoil. Such hope increases the connections among community members as they struggle to build the social components of neighborhood and replenish the social capital.[8] This rebuilding promotes goodwill among returning neighbors and serves as the foundation for community *rebuilding*. We argue that with the rebuilding of "normal" community relations that appropriate pre-Katrina community norms, redevelopment can begin; moreover, programs to rebuild will begin and the higher the degree of connectedness to one's neighborhood, larger community, and city, the more likely redevelopment efforts will succeed. A "timely implementation of post-technological disaster interventions is a challenging endeavor. However, failure to immediately address social impacts of technological disasters exacerbates [the] likelihood of diminished social capital" (Ritchie & Gill, 2007, p. 120). This could introduce impediments to short-term community stability and long-term community recovery and transformation. In summary, "An ecologically and symbolically informed approach to disaster invites us to examine these geopolitical decisions and their interrelationships with such environmental triggers as floods or leaching chemicals and the subsequent interaction of the altered environments with populations and their sociocultural systems" (Kroll-Smith & Couch, 1991, p. 364).

Katrina: Catalyst for Social Change

On January 29, 2006, *The Times-Picayune* noted, "for centuries, canals kept New Orleans dry. Most people never dreamed they would be Mother Nature's Instrument of Destruction" (Marshall, McQuaid, & Scheleifstein, 2006). However, with the abrupt changes that occurred during the hurricane, life as was known prior to Katrina—a culture that thrived on revelry and partying, a city with the motto "*Laissez les bons temps rouler,*"[9] and a haven for the French-Creole traditions and culture—was set on course for change that is currently altering the city. The social changes underway are a mixture of structural changes at the institutional level as new groups emerge to meet the ever-changing needs of the people who demand equal rights, fair treatment, and personal changes. These residents return to the city and must come to terms psychologically and emotionally with the newly configured landscape. With few of the basics, the city lacks running water and electricity, while dilapidated houses and cars still line the streets two years after the passing of the hurricane.

The rebuilding of New Orleans and the Gulf region will demand a change in the usual business atmosphere. With the losses from Hurricane Katrina overshadowing any recent natural disaster, federal disaster agencies and local intergovernmental relationships need to function differently to avoid bureaucratic stumbling blocks that prevent help from reaching the citizens. For example, during the earthquake that destroyed San Francisco, the mayor took control and called on the military and private industries to meet the public's immediate needs. Within days, emergency services were provided, distributing 3,000 tents, requisitioning bedrooms in undamaged houses, and serving millions of meals (despite the fact that the city had forbidden all home cooking for fear of fire). It organized 28 refugee camps located in many of the city's parks, set up field hospitals and dispensaries, organized trainloads of food and clothing, created and maintained telegraph and other communication lines inside and outside the city, and enforced a curfew.

However, the lessons learned about disaster management as a result of the San Francisco earthquake did not transfer to subsequent natural disasters. The Mississippi River flood of 1927 and the Vanport flood in 1948 offer examples of disasters where the sobering lessons in humanity's vulnerability went unheeded and society failed to take measures to understand and lessen the threat to loss of life and property in disaster-prone areas. "A disaster becomes unavoidable in the context of a historically produced pattern of 'vulnerability', evidenced in the location, infrastructure, sociopolitical organization, production and distribution systems, and ideology of society. A society's pattern of vulnerability is a core element of a disaster" (Oliver-Smith & Hoffman, 2002, p. 3).

Social Change and Calamity in an Age of Disaster

How survivors are able to adapt to and reconnect with their environment is at the foundation of social change. In Erickson's (1994) book, *A New Species of Trouble*, he remarks about the disruptive nature of disasters and how they have different effects on individuals and social structures in that they undermine the fabric that binds society. In most cases, these human-induced disasters destroy, in a short span of time, the trust and social capital among citizens and government that developed over years—and in some New Orleans communities, took generations to build. The presence of distrust is distinctive to a risk society[10] in which there is tension and skepticism in reference to trust. In risk societies, there is increasing trust in systems that are supposed to protect society from disasters; however, this trust has produced anxiety and doubt in these systems through historical experience (Ekberg, 2007).

The paradox of trusting the government has historically led to feelings of distrust on the part of certain segments of the population for the protection against disaster destruction that has contributed to a risk society. Moreover, society's disassociation from nature and its lack of use of place-based knowledge contributes to society's reliance on socially constructed systems that fail to cope with natural disaster destruction. "Under certain circumstances, the performance of state-level organizations in the disaster process also became a catalyst for readjusting the character of relationship and interaction between local communities and the structure of larger society" (Oliver-Smith & Hoffman, 2002, p. 10). Understanding the flooding and subsequent social, political, socio-economic, cultural, and human-shaped landscapes is only one aspect of the social change within a risk society.

It is important to remember that alterations in the political and economic landscapes that result from disasters such as Katrina are *manifestations* of social change, not social change itself. People can readily view changes in the political and economic landscape as social change because changes to these landscapes are more easily observed than individual perceptions in a larger society; however, true social change comes when people alter their social interactions on the micro-level so that they are more advantageous to themselves as individuals. Social change, as observed in the political and economic landscapes, then comes when enough individuals have similarly altered their social interactions for self-benefit, and all those who have changed view these changes as truly beneficial. Therefore, as groups, they attempted to persuade the rest of society, through politics and economics, to alter their social interactions so they would be similar to their own.

This is not to say that social change cannot be the *result of* politics or economics; there are numerous historical examples of this occurring. However, when social change is directed from a combination of politically and economically driven forces, long-lasting change is slower and often more resisted. Additionally, this change tends to be temporary and limited to the continuation of political and economic features that drive the social changes; however, when social change is a developmental process, driven by consensus, it is more likely to gain wider acceptance.

Summary

The theories and ideas presented in this chapter underscore the fact that little is known about the process individuals and groups undergo as they navigate the multiple layers of the disaster landscape. Survivors are often confused or angered when they experience landscape lag, which is similar to culture lag. When culture lag occurs, individuals in a society fail to keep up with the rapid pace of technological change, resulting in conflicts among groups in a society when newly developed technology outpaces the skills of the working class. So too can landscape lag result in potential conflicts. When survivors are not a part of the process that brings about change, they lag behind in political, cultural, and economic inclusion.

Although different perspectives offer a glimpse into the transformations occurring in the post-disaster, multi-layered landscape, it is important that the maladaptive patterns are not reproduced or, as Oliver-Smith and Hoffman (2002) maintain:

> While many hazards display their presence quite constantly, others, despite their systemic environmental quality, may not occur with great frequency, allowing the possibility of maladaptive responses over time. If these maladaptive responses become institutionalized, they may lead to an increase in society's vulnerability that in due course may bring about calamity and social collapse. Consequently, hazards and disasters, and how societies fare with them over long periods of time, are potential indices of not only appropriate environmental adaptations, but ideological ones as well. These cultural adaptations include innovation and persistence in memory, cultural history, world views, symbolism, social structural flexibility, religion and the cautionary nature of folklore. (p. 9)

Lefebvre's layered perspective and the ecological–symbolic approach are critical to understanding how survivors adapt and navigate through simultaneous landscapes within the disaster landscape. The landscapes presented in this text, excluding the physical landscape, are not to be reified as the only landscapes to which one can draw social change parallels. In fact, there are other important landscapes we consider as part of the phenomenological interpretation process underscored by the ecological–symbolic perspective. For example, the psychological, emotional, and spiritual landscapes are important in the understanding of social change in the disaster landscape; however, we emphasize the impact of the cultural, economic, and political aspects of the lived experience within the disaster landscape and, in the final chapter, issue a call for a change in the disaster subculture—a change rooted in cultural adaptations that includes a change in world views, offers social structural flexibility, and relies on the historical significance of Katrina.

Notes

1. In 1965, Hurricane Betsy, which was only a category 3 storm, generated a 12-foot wave that came barreling out of Lake Pontchartrain and snapped telephone poles, capsized freighters, and killed more than 50 people (Steinberg, 2006, p. 238).

2. The Chicago Fire (1871); the hurricanes in Galveston (1900), Miami (1930), and Charleston (1989); the Johnstown Flood (1889); the Mississippi floods first affecting Southern cities including Nashville and New Orleans (1927) and then later unsettling Midwestern cities including St. Louis and Des Moines (1993); and the San Francisco earthquakes (1906, 1989). Each disaster caused massive physical damage and large numbers of fatalities and left tens of thousands homeless (Birch, 2006).

3. Population changes occur in (a) the number, (b) the composition, or (c) the distribution of population elements. Cultural changes occur in the content or distribution of culture, that is, changes in (a) the number, (b) the quality, or (c) the distribution of culture traits. Relational changes occur in the relations of (a) individuals or (b) groups to one another. Catastrophic changes occur in the functional adequacy of cultural protections following catastrophes (i.e., the relatively sudden collapses of cultural protections resulting from catastrophes) (Carr, 1932).

4. In An Outline of Social System, Parsons (1961) views society as a system surrounded by three other systems (personality, the organism, and culture). He considers a society in equilibrium when its boundaries with the three other systems are not breached. Thus, social equilibrium consists of the ongoing process of "boundary maintenance": social change that consists of boundary breaking (Vago, 1999). Hence, according to Parsons, "if a sub boundary is broken, resources within the larger system counteract the implicit tendency to structural change" (p. 71). In the Mt. Olivet group example, the larger social structure was in such temporary disarray that adaptive upgrading occurred as a result of internal and external pressures. In essence, social change for classical functional theory is considered as boundary distraction and equilibrium restoration from endogenous and exogenous sources (Parsons).

5. For information regarding household earnings, consult Frymer, Strolovitch, & Warren, 2005. For information regarding health care and healthcare inequalities prior to Hurricane Katrina, consult Atkins and Moy, 2005. For information concerning social and economic isolation and inequality, consult Saenz, 2005.

6. A brief presentation of his theory in Chapter 1 of this text.

7. A. Bocage, NPR Transcrip., Ydstie, 2006.

8. According to the seminal work of Ritchie and Gill (2007, p. 106):

Conceptualization of social capital is concerned with trust, associations, and norms of reciprocity among groups and individuals. According to Putnam (2000), the earliest use of the term social capital was by Hanifan who included "good will, fellowship, sympathy, and social intercourse among the individuals and families who make up a social unit" in his conceptualization (p. 19). Hanifan (1916) suggested an absence or presence of these elements in relationships affects individuals and communities where they live. Others (see Paxton 1999) cite Jacobs (1961) and Loury (1977) as having employed the term prior to introduction into popular use by Bourdieu (1983) and Coleman (1988).

Putnam (2000) defined social capital as "connections among individuals—social networks and the norms of reciprocity and trustworthiness that arise from them" (p. 19). Coleman (1988) contrasts social capital with financial capital, physical capital (e.g., technology), and human capital (e.g., education):

Unlike other forms of capital, social capital inheres in the structure of relations between actors and among actors. It is not lodged either in the actors themselves or in physical implements of production.... If physical capital is wholly tangible, being embodied in observable material form, and human capital is less tangible, being embodied in the skills and knowledge acquired by an individual, social capital is less tangible yet, for it exists in the *relations* among persons. (emphasis in the original) (pp. S98, 100-101)

9. A Cajun expression meaning "Let the good times roll!" that conveys the "joie de vivre" (the joy of living attitude).

10. A risk society is characterized by threats to identity and the risks emerging from the collapse of inherent norms, values, customs, and traditions in addition to dislocation, disintegration, and disorientation associated with the vicissitudes of detraditionalization (Beck, 1992).

Chapter 6
Civic Trust

I will never trust the federal government again.

—New Orleans College student[1]

I believe they do these things intentionally...
so they can flood out those Black neighborhoods...
because every time they have a hurricane,
it always be that way. You know?

—Study Respondent[2]

The ability of a nation's population to trust its government is important in democracies. According to the Panel on Civic Trust and Citizen Responsibility (1999), a democratic government works best when the governmental structure is the incorporating element that links a network of various communities, associations, and families and contains citizens who care and have a reasonable amount of faith in their official leaders. Although full faith in any government structure is most desired, human history has shown that no matter how popular the leader or government structure, inequalities and injustices exist and breed distrust (Aberbach & Walker, 1970). Theoretically, democracies attempt to decrease the overall distrust of citizens by appointing individuals to positions of power who will advance the majority of citizens' interests; "thus trust on the behalf of citizens toward a representative depends on the constituent's perception that the representative shares his or her interests" (Ruscio, 1996, as cited in Miller & Rivera, 2006a, p. 40). In the event there is an existence of distrustful citizens who are convinced that the government only serves the interests of a few rather than the interest of all, a barrier is created that impedes the realization of the democratic ideal (Aberbach & Walker, 1970). Moreover, certain factors, such as age, education, income, political alienation, feelings of deprivation, expectations about treatment from government officials, strength of drive to self-assertion, and race, can influence the level of distrust inherent in a population.

With a front row seat to the abrupt changes in the physical landscape and the devastation of their homes, communities, and city, ordinary citizens are able to see the United States government in action, or in many cases inaction, and many of them do not like what they see (Gunter & Kroll-Smith, 2007). What the citizens saw was a violation of the basic assumptions they hold as citizens in a democracy. Most important is the fiduciary responsibility a governing nation has for the protection of its citizens. Gunter and Kroll-Smith (2007) maintain as follows:

> [when] faced with a recalcitrant and worrisome environmental problem, people are likely to find government agencies unhelpful, perhaps hostile, or perhaps simply uninterested. Indeed, they might find, or perceive. . .that instead of protecting them from harm, government actions put them at greater risk. Instead of making decisions in a fair and impartial manner, public officials are apt to be viewed as siding with the wealthy and politically connected. (p. 71)

Hurricane Katrina challenged the basic assumptions of trust that the American people held in the institutions, organizations, technology, and biophysical environment. Katrina also highlighted decades of inequality and injustices regarding ethnic minorities and low-income individuals that were prevalent in New Orleans. American and local history[3] are littered with a litany of responses that engender citizens' frustration. Frustration runs the gamut from lies and cover-ups to foot-dragging, indecisiveness, and dodging responsibility; these actions all point to a pattern of neglect that leaves citizens feeling betrayed and disillusioned, resulting in a loss of trust (Gunter & Kroll-Smith, 2007). Moreover, the political and socio-cultural landscapes that were prevalent prior to Katrina and emerged in the storm's aftermath reinforced a sense of distrust that certain segments of the population had harbored for decades toward the local, state, and federal government. In essence, "the New Orleans disaster captures on a large scale the downward spiral of disintegrating relationships between residents and officials; government, it seems, often fails to respond to local environmental controversies and catastrophes in the manner citizens expect" (Gunter & Kroll-Smith, 2007, p. 70).

Understanding how to rebuild and foster more civic trust in organizations, institutions, and technology is a matter that must be addressed to avoid duplicating the political, social, and cultural landscapes that were inherent to the city prior to the storm if New Orleans is to be rebuilt. The only thing that will foster an environment attractive enough for citizens to want to return and for others to want to settle there for the first time is trust in the political institutions and organizations, such as the offices of the mayor and governor and the organizations and agencies responsible for the public safety of all people, in addition to the mitigation technologies, such as levees and pumps. This chapter will discuss the importance and dynamics of trust and its effect on civic engagement and social capital that are necessary to develop when building strong social, cultural, and political landscapes.

Trust

For any human organization to function in a coherent manner, there must be some level of trust that has been developed between the individuals. According to Hardin (2002), trust is always relational. Trust between individuals depends on their relationship, which is either developed directly through their *ongoing* interactions or indirectly through intermediaries and reputational characteristics. Trust can be the result of incentives to develop and maintain relationships between individuals.

> The trusted party has incentive to be trustworthy, incentive that is grounded in the value of maintaining the relationship into the future. That is, I trust you because your interest encapsulated mine, which is to say that you have an interest in fulfilling my trust. (Hardin, 2002, p. 3).

The incentives to trust one another for mutual benefit provoke the development of expectations between the two individuals who base their understanding of the other's interests in relation to themselves (Hardin, 2002). Although beneficiary incentives entice individuals or groups involved in a relationship to more easily trust one another, there is always a concern that trusting can detrimentally affect one's interests. This risk is manifested in the possibility that the other actor being trusted will abuse their power of discretion to the disadvantage of the trustee (Hardin). However, because individuals and groups have the ability to alter their interests at any given moment, trusting in another entity is inherently latent with risk.

Trust and the Political Landscape

A population's ability to trust its government is fundamental to a democratic regime's ability to make legitimized decisions. The notions of popular sovereignty and majoritarianism have been provided as forms of democracy that seem to legitimize a democratic regime's decisions and actions. Although popular sovereignty and majoritarianism are forms of democracy, they have the ability to be antidemocratic because they tend to neglect the interests and needs of the minority (Post, 2006). Moreover, these versions of democracy have historically led to the rule of dictators who carry out the genuine and spontaneous will of the population and alter the electorate standards, which has led to the election of monarchs and other forms of "dynastic" type leadership into office[4] (Post, 1998a, 1998b, 2006). In both of these versions of democracy, there is a distinct level of trust in government; however, in reference to popular sovereignty, trust may vary to any degree at any given time, and in majoritarianism, the population's trust is only of consequence if it varies in respect to the majority.

The political landscape that developed in New Orleans led to a system of democracy that neglected the interests of the majority (the lower socio-economic group) and deferred to the interests of the minority, bringing into question why political decisions were historically made in the manner they were. According to Michelman (1998), a true democracy is distinct from popular sovereignty and majoritarianism because democracy is a normative idea that refers to substantive political values whereas popular sovereignty and majoritarianism refer to the ways in which decisions are made by a government.[5] Post (2006) maintains that democracy should not be confused with decision-making systems and procedures but rather understood as an identification of core values that the government seeks to internalize in its citizenry. If Post (2006) is correct in his reasoning, it raises the question of what values the City of New Orleans was attempting to instill in its marginalized, low-income, minority neighborhoods.

The most prevalent value in democratic governments is the idea of self-determination (Kelsen, 1961). Self-determination is often interpreted to mean that individuals are responsible for governmental decisions and actions, either by making decisions directly or by electing people who do (Meiklejohn, 1948). Post (2006) takes this idea further by arguing that self-determination requires that people have the warranted conviction that they are engaged in the process of government themselves. Without the presence of self-determination in individuals or communities, distrust can develop. Yamagishi (2001) asserts that distrustful individuals and groups usually develop because they have been socially isolated and, therefore, do not have the same opportunities for developing their own social intelligence as other people have had. This leads to a lack of internalization of self-determination on the part of these groups, thereby decreasing their belief that the system is concerned with their interests and needs.

Negative self-determination perspectives are also produced through the occurrence of historic events that erode the trust Americans have in various organizations and institutions (Welch et al., 2005). Welch et al. (2005) explains that, whether justified or not, public perceptions about voting were challenged in the presidential election of 2000, provoking doubts about the integrity of the electoral process. Questioning the integrity of the electoral process challenged individual attitudes of self-determination in reference to electing individuals that shared the public's interests, which if protracted over an extended period of time, could subvert any remaining faith that Americans may have in the democratic process (Welch et al., 2005). In New Orleans, the lack of initiative on behalf of the city's levee boards to alter its residents' vulnerability to natural hazards, or even to incorporate citizens' perspectives on disasters, directly affected the citizenry's sense of self-determination. Moreover, the city's social and cultural landscapes led to the marginalization of minorities and low-income residents, which further affected these citizens' sense of self-determination, leading them to feel as though they had little control over their future.

The individual realization of self-determination in segments of the population can also be affected by the presence of inequalities among different segments of a population. Aberbach and Walker (1970) contend that segments of a population will always feel as though inequities exist and that these inequities premise distrust on the behalf of these groups toward the government. Inequalities have the ability to depress civic participation, whether directly or indirectly, through their effects on trust.

> where inequality is higher, the poor may feel powerless. They will perceive that their views are not represented in the political system and they will opt out of civic engagement. Second, trust in others rests on a foundation of economic equality. Where resources are distributed inequitably, people at the top and the bottom will not see each other as facing a shared fate. Therefore, they will have less reason to trust people of different backgrounds. (Uslaner & Brown, 2005, p. 869)

As described by Meiklejohn (1948), when disgruntled and unsatisfied groups opt out of civic engagement, they are also opting out of their responsibility for governmental decisions and actions. Because of the presence of the Protestant work ethic in American society, when people or groups are unable to gain access to resources through "hard work," they internalize their lack of resources as a failure in life (Avery, 2006). This internalization of failure perpetuates distrust in the government when enough begin to believe that their failure to acquire resources is not because of a lack of "hard work" but because the system is structured in a way that is specifically inequitable to their social group (Avery, 2006). Also, the notion that people with varying access to resources perceive others as not facing the same shared fate leads to a disassociation between these different segments of the population, which results in less trust between the groups. Deutsch (1960, as cited in Cook & Cooper, 2003) contends that for people to cooperate with one another, they must develop a sense of mutual trust, which is advantageous when individuals or groups are oriented to one another's welfare. When a lack of a sense of shared fate exists among different groups, individuals in those groups show signs of particularized trust in that they only trust and engage in relationships with others similar to them (Uslaner, 2001; Uslaner & Brown, 2005; Wuthnow, 1998). Additionally, trust not only facilitates cooperation, but it also contributes to the maintenance of social order (Cook & Cooper, 2003; Luhmann, 1988; Putnam, 1993, 2000). Individuals and groups who are distrustful of the government are more inclined to take radical and socially destabilizing actions than they would if trust were present.

The presence and perception of social inequities and injustices in a society promote distrust and feelings of paranoia among the disadvantaged groups. Colby (1981, as cited in Kramer, 2004) noted that paranoia occurs through beliefs in ideas of being "harassed, threatened, harmed, subjugated, persecuted, accused, mistreated, wronged, tormented, disparaged, vilified, and so on, by malevolent others, either specific individuals or groups" (p. 140). Although Colby (1981, as cited in Kramer, 2004) defines these beliefs to be false from an individual's psychological standpoint when they are justified through historical experience,[6] it is expected that social groups develop a collective paranoia, which further develops into a collective "them against us" mentality. This mentality can develop among members of each group from both in-group and out-group participants who view the individuals of alternative groups as all the same, which Kramer (2004) defined as the out-group unitization hypothesis:

> Out-group unitization reflects the tendency to differentiate less among out-group members than among in-group members when doing the social auditing. As a consequence, a breach of trust from one member of the out-group can be repaid to any other member. If an out-group member insults one, one can retaliate against any other member of the out-group. The target of the retaliation naturally perceives the retaliatory act as gratuitous aggression, further enhancing distrust and suspicion [among the group]. [Alternatively], if an in-group member insults one, one retaliates directly against that person. (p. 141)

Kramer (2004) also noted that out-group unitization can work inversely, with positive responses from both in-groups and out-groups; however, when positive actions are made, they are individually perceived, which limits the positive action's effect on the group's perception of the other. In-group and out-group conflict can take place over time and be facilitated by institutional internalization, especially in reference to group differentiation based on racial characteristics.

> Institutions do not simply have momentary breakdowns during racial crises; they create these crises by structuring U.S. politics in ways that enable the maintenance of racial inequality. Consequently, blaming individuals or even temporary institutional failure does not go far enough in helping us to understand why it was that a disproportionate number of the poor in New Orleans were African American, or that African Americans in New Orleans were disproportionately poor. What we watched unfold on our television sets was not an accident, but the institutionalized result of centuries of concerted decision making by political actors at the local, state, and national levels, going back to the days of slavery and continuing up to our current political moment. (Frymer, Strolovitch, & Warren, 2006, pp. 46-47)

In this regard, institutions of government actively attempt to foster the collective paranoia of specific groups in an effort to stimulate defensive non-cooperation. In specific reference to how collective paranoia plays a part in civic trust and participation, groups that are collectively paranoid may participate in the behavior of defensive non-cooperation (Kramer & Brewer, 1986). In the context of civic engagement, defensive non-cooperation takes the form of opting out of voting in an election because the group has the expectation that their interests are not going to be addressed, or worse, are going to be violated by the incumbents regardless of who is elected. Therefore, the group opts out of participation because it does not trust that the process will yield beneficial results.

Typically, defensive non-cooperation does not formally manifest itself among specific social groups; however, this trend does appear among individuals. According to some research, certain individuals within specific social groups are more prone to this course of action than others, due in part to their social group's collective paranoia[7] in relation to institutions that intend to aid them.[8] This defense mechanism has a two-fold result in relation to trust. First, it illustrates to outsiders that they are unwilling to fully engage in cooperation because their paranoia limits them from truly trusting others to do their part in pursuit of goals. Second, this type of action provides an excuse to outsiders not to bring them into discussions and cooperative actions because they have set a precedent that illustrates they do not want to engage in cooperative actions or, if they do, there is a heightened level of potential conflict that may occur, which may deter the development of social capital from ever occurring.

Alternatively, some researchers believe that in democratic environments, it is the lack of trust that fosters dissatisfied individuals and groups to become involved in government. An example follows:

> Political life is extremely confrontational. The goal of politics is to win, to defeat the opposition, and has become even more so in this era of heightened partisanship and loud voices. Whereas civic engagement depends on trust, political action thrives on mistrust (Warren, 1996). People will be more likely to get involved in political life when they get mad and believe that some others, be they people or political leaders, cannot be trusted. When people are upset, they are more likely to take direct action in their communities. (Dahl, 1961; Scott, 1985, as cited in Uslaner & Brown, 2005, p. 875)[9]

Avery (2006) contends that lower levels of political trust among certain social groups, specifically African Americans, induced by discrimination and inequality, foster government participation. This is most apparent in the American political system, where the incorporation of African Americans and other racial and ethnic groups has been given a full promise to American political, social, and economic life; this promise has been extremely challenging to fulfill to any equitable degree, as can be seen in New Orleans (Lieberman, 2006). For Avery (2006), mistrust on the behalf of minority social groups and its affect on governmental engagement stems not from unhappiness or distrust in a specific political regime, its policies, and its leaders, but from the entire political system that continues to marginalize these social groups regardless of regime or policies. In this vein, Rosenstone and Hansen (1993) explained that trusting citizens are not more likely to engage in political activities and are, in actuality, less likely to be interested in politics because they are already content. What induces political engagement today is the emergence of the contemporary civil society in America:

> It is the gradual disappearance of safe streets, stable families, secure employment, and the enduring relationships with relatives, neighbors, merchants and co-workers that make an ordered life possible. It is the unraveling of the strands of community—of what many are now calling civil society. (Wuthnow, 2002, p. 61)

Distrust of government throughout history has facilitated, and in some cases encouraged, the breakdown of relationships that fosters an environment for political engagement and an atmosphere of trust.

Historic Distrust

For some groups in American society, government institutions have continually engaged in decisions and actions that have disproportionately been to the groups' disadvantage (Frymer et al., 2006; Lieberman, 2006). These decisions to instill certain values, such as the Protestant work ethic, into the minds of the entire citizenry, when structural barriers limit some no matter how hard they try, further produce distrust toward the government. The Protestant work ethic can serve several functions:

> not only epistemic needs for understanding, meaning, simplification, and prediction of one's environment, but also social and personal needs for controlling one's environment, supporting values, warding off perceived threats, and *maintaining social relationships.* (Wegener & Petty, 1998, as cited in Levy, Freitas, Mendoza-Denton, & Kugelmass, 2006, p. 76)[10]

Values in a society maintain certain social relationships that particular groups continually have a difficult time buying into. According to Levy et al. (2006), Katrina challenged the value of the Protestant work ethic among Blacks: the researchers observed that fair treatment is not guaranteed, even for those people who work hard. The challenge of this valued belief created a reduction in perceived trust in government by Blacks, which was also a reenactment of how this same group generally felt in reaction to prior mass floods in the region.

Bullard (1990a, 1990b, 2000) contends that governments create sacrificial zones in which everything tied to a specific place and time is expendable and has little value. For many people in the South, it is hard to escape the remnants of the former plantation system that exploited both the people and the land and has become the nations' dumping ground for toxic wastes (Bullard, 2000). Moreover, it seems from historical experience that new zones of sacrifice emerge whenever Blacks settle in great numbers in areas where they were not inherently welcome or where there have been years of discriminatory sentiment (Rivera & Miller, 2007a). Blacks were displaced by flooding in 1927 due to inadequate government mitigation and response. Many migrated to other areas of the nation, where they were similarly neglected and sacrificed by local and federal officials. Government neglect resulted in public policy decisions that displaced hundreds of thousands of Blacks and left thousands of others to die. During the 1927 Mississippi Flood, Blacks and the land they lived on in New Orleans was sacrificed to save the rest of the city:

> To save New Orleans, the leaders proposed a radical plan. South of the city, [where] the population [is] mostly rural and mostly poor, the leaders appealed to the federal government to essentially sacrifice those parishes by blowing up an earthen levee and diverting the water into marshland. They promised restitution to people who lost their homes. Government officials, including Commerce Secretary Herbert Hoover, signed off.
>
> On April 29, the levee at Caernarvon, 13 miles north of New Orleans, succumbed to 39 tons of dynamite. The river rushed through at 250,000 cubic feet per second. New Orleans was saved, but the misery of the flooded parishes had only started. The city fathers took years to make good on their promise, and few residents ever saw any compensation at all. (Slivka, 2005, p. 26)

Although Rivera and Miller (2007a) only described the Mississippi Flood of 1927, the 1948 Vanport Flood in Oregon,[11] and Hurricane Katrina, there are many similar experiences that have occurred over time and across the entire nation in reference to different ethnic minorities that have had detrimental effects on these groups' trust in the government. Moreover, following Hurricane Katrina, many Black residents thought that the government had repeated what was done in 1927—they destroyed a levee to save the more affluent segments of the city, illustrating the expectations of distrust among the population toward the government.

> Katrina revealed not only how white America historically has felt about black cities and spaces—viewed more like colonial outposts that require a constant military presence—but also how such racist ideologies have translated into forms of underinvestment, criminalization, and a brutal lack of compassion on the part of the racial state. (Giroux, 2006, p. 27)

In addition to historically sacrificing certain segments of the population, modern disaster recovery policy has generated distrust toward the government. As presented in Chapter 4, the federal government has crafted legislation that both fosters growth into geographic areas prone to disaster by offering insurance programs and other mechanisms to replace destroyed property and attempts to ward people and businesses away from settling into vulnerable areas. The federal government has also crafted legislation that leaves local governmental authorities responsible for the immediate response, relief, and mitigation of disasters; this responsibility fell into the hands of irresponsible elected officials and corrupt levee boards, as was discussed in Chapter 4. "The legislation that was used to deal with Hurricane Katrina also left all mitigation efforts to the local governmental units for implementation and development" (Rivera & Miller, 2006a, p. 517).

> Furthermore, through Rivera and Miller's (2006) analysis of past mitigation and relief policy, the federal government's tendency to let local governments be responsible left the people of New Orleans and the Gulf Coast at the political benevolence of governmental authorities, which viewed socially vulnerable communities of the region not significant enough to warrant mitigation and relief plans. (Rivera & Miller, 2007a, pp. 517-518)

Moreover, even in the event of a disaster, relief mechanisms available to local governments are not guaranteed to occur at all or in any hurried manner, as first assumed by Congress in 1950[12] and vividly portrayed by the actions taken, or not taken, by the different levels of government during the aftermath of Hurricane Katrina. These historical experiences have created distrust among Blacks, particularly in relation to the government, which has affected their ability to develop trust in government institutions (Paxton, 1999).[13]

Social Capital and Civic Participation

Selznick (2002) contends that, in a "deliberative democracy," individuals and social groups can interact with one another to make the best decisions about a society. A deliberative democracy allows people in a community to reason with one another and act collectively. This is not to say that deliberative democracy simply allows all perspectives to be voiced on an issue for the sake of voicing them, but rather that by voicing differing perceptions and preferences deliberation leads to the criticism of perspectives, which in turn leads to collective intelligence (Selznick, 2002). The flow of information is the first step in building social capital among a community.

In this context, social capital is defined as an investment in social relationships for expected returns in the marketplace or, rather, the larger society (Lin, 2001).[14] For Lin, social capital enhances the outcomes of actions due to four factors: information, influence, social credentials, and reinforcement. These factors explain that the flow of information allows people and groups to be exposed to opportunities and alternate points of view that may not have been available otherwise. The social ties developed between groups and individuals have the ability to exert influence on the agents who are critical in making decisions. In some cases, social ties can develop between the individuals or groups with some of the agents that make decisions. Due to the social ties that can develop among these decision-making agents, they can use their position (i.e., authority and supervisory capacities) to influence other decision-making agents, carrying more weight than just one individual. A social tie between individuals or groups attests to their social credentials or, rather, to their access to resources through social networks and relations. Lastly, social ties reinforce identity and recognition among those who have developed a relationship. Those groups or individuals who share similar interests and resources provide emotional support to those with resources.[15]

Although social capital may focus on the development of relationships among individuals, Bourdieu views social capital from a group perspective; specifically, Bourdieu's works (1980, 1983/1986) investigate how certain groups develop and more or less maintain social capital as a collective asset and how much this collective asset enhances the life chances of the group's members. From this perspective, the amount of social capital a group has is reliant on the size of a group's connections and on the volume of capital that these connections possess (Bourdieu, 1980, 1983/1986). Social capital is the result of repeated exchanges reinforcing mutual recognition and boundaries to affirm the collectivity of capital and each group member's claim to it. Moreover, social capital is shared by a specifically defined group that contains clear boundaries, obligations of exchange, and mutual recognition (Bourdieu, 1980, 1983/1986). Scholars push this idea further by arguing that the number of social associations and their participation in society indicate the extent of social capital within a society.

> These associations and participation promote and enhance collective norms and trust, which are central to the production and maintenance of the collective well-being. (Putnam, 1993, 1995, as cited in Lin, 2001, p. 23)

If social capital is the investment in relationships to access and borrow the resources of others for the betterment of the group (Coleman, 1990) and if the degree to which people or associations interact with one another gauges the level of social capital in a society (Putnam, 1993, 1995), then what resources are sought after by individuals or groups? In the case of civic participation, social groups are investing in social capital to gain resources through networks, which may take a tangible form such as finances, but they predominantly seek less tangible resources, such as political power.

In the American political system, however, the expectation is that wealth brings political power, and conversely, if one has political power, they also have wealth. Due to this common notion, those with the resource of wealth or political power tend to have higher standing in relation to other individuals; therefore, they are given opportunities to make decisions on behalf of or in the name of the collective group (Lin, 2001). Alternatively, Lin (2001) maintains that individuals with less valued resources have lower standings in the group and are more prone to structural constraints and less opportunities to engage in the decision-making process. Because this dichotomy exists, the individuals or groups with more resources attempt force group consensus because there is incentive for them to sustain and promote their standing in the group. These already resource-rich individuals or groups advance their standing further by gaining more valued resources or by manipulating the group's value consensus so the resources they possess continue to have high value (Lin, 2001). Lin's (2001) perspective suggests that when a threat emerges toward the group in this type of system, those with more resources (who are also the decision makers) tend to survive the threat, whereas those with fewer resources are impacted more by the threat and sometimes do not survive.

Social Capital and Communities

If social capital develops within a community context, there must be an understanding of what a community *is*. Although there is some variance in the definition of what constitutes a community, Wood and Judikis (2002, p. 12) put forth the following six elements:

1) a sense of common purpose(s) or interest(s) among members;
2) an assumption of mutual responsibility;
3) acknowledgement (at least among members) of interconnectedness;
4) mutual respect for individual differences;
5) mutual commitment to the well-being of each other; and
6) commitment by the members to the integrity and well-being of the group, that is, the community itself.

By this definition, a community is not limited to any specific geographic area and could, in theory, be composed of members spatially distributed over large distances. For example, ethnic and racial groups could be communities, even if they reside in different sections of the United States, as long as they are working toward the well-being of the entire group with some manner of common interest, interconnectedness, and mutual respect for one another (Wood & Judikis, 2002). However, although an outsider can view a group as a community because it may appear to exhibit the six elements, it is important to know how the group defines its own structure when defining a community (Denhardt & Glaser, 1999).

A community's concept of itself or how it defines itself is referred to as the community's orientation, which is generally produced by its social cohesion. This includes its history or experience with a level of commitment to organization and a willingness to commit to helping improve the group. Secondly, community orientation is expressed through its own leadership and willingness to cooperate with the leadership of the larger "outside" societal community (Denhardt & Glaser, 1999). Although Denhardt and Glaser (1999) discussed communities in reference to specific geographical areas, their definition does not restrict a community to geographic location. It is easy to apply these definitions of community to New Orleans residents prior to the storm. Furthermore, by combining Wood and Judikis' (2002) and Denhardt and Glaser's (1999) definitions, the displaced portions of the New Orleans population can also be considered a community. The displaced people, separated by vast distance, still exhibit elements of a connected community even though they are no longer in one geographic area; however, these survivors in diaspora are gradually assimilating into other communities.

A community's orientation is important and depends on its epistemological and ontological predispositions (Dixon & Dogan, 2003). According to Hollis

(1994), epistemological predispositions relate to people's contentions about what is knowable, how it can be known, and the standard by which the truth can be judged. In this way, social knowledge rests on interpretations made from cultural practice, discourse, and language, which are generated by acts of ideation that rest on inter-subjectively shared symbols that allow the reciprocity of perspectives (Dixon, Dogan, & Sanderson, 2005).[16] In other words, groups or communities must internally discuss and assign the value of truth to phenomena, whether they are natural or institutionally implemented. The emphasis is not only on the validity of the phenomenon itself but also on the actual motivations behind the event. In reference to New Orleans, communities across the nation assessed the validity of the government's reasons for a slow and ineffective response to Katrina as false. Although there were several factual reasons for why governmental response efforts were lacking, a large segment of the community devalued the government's reasons because they considered the truth to be that the slow relief efforts were motivated more by racial sentiments harbored by officials than an inadequate infrastructure based on cultural and historical experience (Dyson, 2006). This construction of what the community viewed as truth was also motivated by its ontological predispositions.

As proposed by Hollis (1994) and Dixon et al. (2005), ontological predispositions relate to people's contentions about the nature of being, what can and does exist, what their conditions or existence might be, and to what phenomena causal capacity might be ascribed. These predispositions can be based on structural propositions that social structures themselves impose on a group or community (Baert, 1998). This notion supports the argument made by Frymer et al. (2006) that institutions perpetuate the existence of inequality among certain groups or communities. Ontological predispositions have a significant influence on a community's general inclination to trust and, subsequently, to develop formal social capital with governmental officials and institutions.[17] MacGillivary and Walker (2000) contended that although formal social capital is influential at the local community level, it is more important at the national level, where social capital is the culmination of all communities and their levels of trust. Moreover, the belief that social structures exert their values on communities over the values of the specific communities retards civic participation on behalf of the individual and, subsequently, the community because they feel insignificant.

> Social structures are regarded as constraining in the way they mould people's actions and thoughts, and in that it is difficult, if not impossible, for one person to transform these structures. (Baert, 1998, as cited in Dixon et al., 2005, p. 6)

Ontological predispositions are important in building social capital among communities because they indicate the level of trust inherent in a community.

Social capital and trust are further complicated by the fact that individuals belong to multiple communities and reside in differing social, cultural, and political landscapes. Wood and Judikis (2002) present the multiple communities theory, which proposes the following:

> that (1) every adult holds simultaneous membership in several communities; (2) each community exercises influence on individual needs, perceptions, values,

attributes, and behaviors; (3) the influences of the different communities are sometimes harmonious and reinforcing, sometimes disharmonious and conflicting; (4) behavior by individuals (both public and private behavior) can be understood in terms of collective experiences and influences within and across communities; (5) the similarities and contradictions in the influences of communities, as well as the specific nature of those influences, are important in determining and understanding the values, perceptions, attitudes, and behavior of individuals; and (6) relationships between communities affect the roles people play within communities and across communities. (p. 30)

Wood and Judikis' (2002) theory complicates trust and social capital because it implies that although individuals belong to a broad community, they also belong to other communities that influence their perceptions and evaluation of events. Therefore, people alter their epistemological and ontological predispositions over time in relation to the broader community. According to Wood and Judikis (2002), individuals alter their perceptions momentarily because of the community's needs, values, perceptions, actions, and norms that develop over time. These community alterations therefore alter individuals' perceptions, which subsequently alter the community's perceptions, leading to new alterations of individual perspectives. This cycle of changing perception, which is a direct result of both a dynamic social system and the fact that every individual belongs to multiple communities, makes it more important that the broader society and community develop social capital and trust so that society does not break down into incompatible communities.

Katrina altered local residents' belief that the government would be able to protect them from harm and that if they were not protected initially through mitigation projects, they would be aided later. Because this perception was distorted by both the lack of protection of mitigation projects and the fact the government did not respond quickly, survivors' perceptions, needs, values, and norms were altered to cope with the day-to-day struggle to survive. These alterations, although motivated by a specific event in time, will change the community's perceptions, beliefs, values, needs, and level of trust in the future, and they will never resemble what they were prior to the storm. Although the community will not be incompatible with other communities to the extent that it cannot work with them to survive, the development of social capital will be retarded because of the new perceptions that were created (and in some circumstances reinforced).

When institutions fail to perform duties that ensure public safety, the social context of that failure is viewed as either an act of trustworthiness or a betrayal of trust. Freudenburg (1993) coined the term *social recreancy*, which refers to the contempt for the failing institutional actors and the anger felt toward the failing institution in general on the behalf of the disillusioned population. Freudenburg (1993, as cited in Miller, Gill, & Picou, 2000) argued the following:

> When a failure to perform specific behaviors occur, those citizens assessing the institutional actors are less likely to depend on these specialized individuals representing institutional actors, thus rendering the individual and institutional actors less credible. It is difficult to separate trust or a lack of trust in an individual or a specific institution; rather, the citizens of a community are likely to

blame the entire structure that encompasses trust, agency, responsibility, or other expected obligations of the collectivity. (p. 15)

In this way, the failure of the individual representatives of a governmental agency or the entire governmental structure effects the trust citizens have in the entire system. In the case of Katrina, failures to respond effectively and mitigate the disaster caused citizens to question the trustworthiness of the entire government and, more specifically, the Federal Emergency Management Agency (FEMA). Social recreancy developed toward both President Bush, as the representative of all government actions and decisions, and Michael Brown, as the person in charge of FEMA during Katrina. The management of the crisis by these two individuals resulted in a decrease in the public's trust that neither FEMA nor the federal government, in general, would respond to any disaster event or come to the aid of its citizens. Although it is arguable that FEMA can or cannot respond to other types of disasters and that the federal government has little regard for some segments of the population, the point is that the citizens now perceive FEMA and the federal government as less trustworthy than they were prior to the storm.

Katrina's effect on New Orleans' residents' formal social capital and trust was evident in the storm's aftermath, in which corrosive communities (Miller, 2006b) and fragmented communities (Denhardt & Glaser, 1999) manifested themselves. A corrosive community can be characterized by social disruption, a lack of consensus about environmental degradation, and general uncertainty. The manifestation of a corrosive community directly affects trust and the ability of social capital to develop. Ritchie (2004) explained that social capital is degraded in a corrosive community; moreover, trust between individuals and organizations also degrades, which is directly due to the presence of a sense of uncertainty. Additionally, a fragmented community exhibits few signs of community identification, and members act purely out of self-interest for their survival (Denhardt & Glaser, 1999)). In such an unstable social setting, trust cannot develop either between individuals or between individuals and organizations, especially if the cause for social disruption is attributed to the actions of leaders and institutions. In social settings and communities such as these, it is more probable that negative manifestations emerge, such as ethnocentrism and corruption (Ritchie & Gill, 2007). This was exhibited by the external communities surrounding New Orleans when they stopped evacuees from entering their towns. Although actions such as these, and the manifestation of corrosive and fragmented communities, can result from an extremely devastating disaster that alters all environmental and social norms, it is more likely that they exert themselves after these events because the community as a whole was not a "healthy community" to begin with, as was the case with New Orleans.[18]

The development of corrosive and fragmented communities can be fostered by outside forces, such as the media. Traditionally, the mass media has both been thanked and blamed for many of the perceptions and beliefs held by members of the general public (Mehta, 1995). In terms of the processing of information and blame assignment to governmental entities, the media plays a more vital role than just providing bits of information. These news agencies set the agendas and construct public attitudes that encourage both consumption and citizens' perceptions of environmental risks and other social groups. In the case of Katrina, the news media depicted scenes of Blacks "looting," whereas the same actions taken by Whites were depicted as subsistence survival. These depictions fostered an environment that generated fear, uncertainty, and mistrust in the minds of all of the groups involved in the struggle for survival, depressing the ability of all groups to redevelop social capital. A lack of trust leads to individuals and communities feeling alienated and powerless, contributing to the potential of fragmented and corrosive communities developing.

Disaster relocation further degrades social capital because, although a new location may place an individual or group in a more stable physical environment, the dislocation places them in a more socially unstable environment. McCully (1996) maintained that people who are dislocated have historically had little option but to migrate to urban environments and typically into urban slums. These resettled people, although placed in a stable physical environment, face new challenges of adequate resources, various degrees of social disarticulation, and marginalization by host communities. This development affects not only the physical and social security of these uprooted individuals and groups, but also the confidence in their culture and social fabric (Oliver-Smith, 2005). In addition to being dislocated, the settlement of new populations into a disaster area would likewise affect these new populations' social capital development.

As survivors return and new populations begin settling into areas with extremely limited resources, the ability of these groups to develop formal and informal social capital is hampered. Dislocation and social disruption point to the ineffectiveness of human and institutional efforts to make life reasonably predictable for a nation or city's population (Oliver-Smith, 2005), which are also violations of citizens' expectations of the government. This violation of expectations on behalf of citizens regarding their government again degrades trust and social capital, especially formal social capital. Kroll-Smith and Couch (1993a) maintain that community trauma that is the result of a disaster creates a collective stress that includes cultural change involving reality disjuncture (a lack of shared group assumptions) and structural change, which disrupts social routines and networks. Along with prolonged material deprivation (Dirks, 1980) and exposure to rapid cultural and structural changes comes an erosion of basic identities and interactions within a community.

Trust, Betrayal, and Consequences in the Aftermath of Hurricane Katrina

Trust and the ability to reestablish relationships among residents returning to the city is only one facet of the road to recovery within the cultural, economic, and political landscapes. On many levels, trust among citizens diminished as residents experienced or viewed news reports of the calamity befalling the city. Hurricane Katrina exposed and made real the ugly underbelly of the government's inactivity toward segments of the population whose voices have traditionally remained unheard. Instead of acting in ways that promoted effective governance and enhanced social capital, the actions of the state, local, and federal government failed to respond to concerns regarding the catastrophe as well as ongoing issues of housing, crime, and equitable rebuilding with compassion and sensitivity, which created feelings of disillusionment, distrust, and betrayal.

For Gunter and Kroll-Smith (2007), the loss of trust can take three forms when public officials fail to act in good faith and betray the basic assumptions of civic trust. These assumptions of civic trust root communities historically, socially, and emotionally to a place. Gunter and Kroll-Smith's (2007) typology for the betrayal of trust views the complex array of factors that take into account the intentions and perceptions in which social actors engage. They use the terms *premeditated betrayal*, *structural betrayal*, and *equivocal betrayal*.

Premeditated betrayal refers to the reckless disregard demonstrated by public officials. It is the most egregious example of the loss of collective trust among a citizenry who all experience what Erikson (1994) calls a collective trauma. For example, when the head of FEMA and members of the Bush Administration stated that they were unaware of the conditions in New Orleans, it contradicted the fact that news media, internal documentation, and pre-disaster briefing notes showed the predictions and warnings well in advance of Katrina's landfall. Not only did the public see the masses struggling for safety, they also saw an inattentive President who remained on vacation at his ranch in Crawford, Texas, instead of returning to engage advisors. The news media further emphasized that other senior Bush administration officials were also on vacation during Katrina's initial landfall. Secretary of State Condoleezza Rice was in New York City shopping and attending Broadway shows, and Vice President Richard Cheney was also on vacation in Jackson, Wyoming; two senior Cabinet members were perceived as inattentive to the needs of Gulf Coast residents. When the crisis in New Orleans became a national emergency, the message was that the Bush administration did not care about the plight of those forced to fend for themselves.

Other acts by local politicians, such as the timing of the mandatory evacuation orders or positioning the needed transportation around the city so citizens could leave the city in an orderly manner, were done well in advance of harm. Despite the advanced warning, there was no system in place to coordinate the mass scale volunteering and emergency personnel needed for such a large disaster. Although many officials have repeatedly stated that no one expected such a disaster to occur or that the unimaginable caught America off guard, denying knowledge of the impact a disaster such as Katrina could have had in the area is dishonest. Information was presented in a 2001 FEMA report that stated a hurricane blow to New Orleans was one of the three most likely mega-disasters to occur in America.

> barely a year before the prediction [of Hurricane Katrina] came true, a gathering of state, local and federal officials used computer models to examine the impacts of a fictional "Hurricane Pam." That exercise showed city levels collapsing from a surge tide arriving almost unimpeded from the Gulf. It showed "the bowl" of New Orleans catastrophically filling with water. It showed a million people evacuating the region and half a million buildings destroyed. It implied thousands of fatalities from drowning and subsequent hardships. . . . And

yet on September 1, 2005, three days after Hurricane Katrina hit, President Bush stood before reporters and said, "I don't think anyone anticipated the breach in the levees." (Tidwell, 2006, p. 30)

The second form of betrayal is structural, which occurs when organizations fail to fulfill their responsibilities without any intent to do so. Structural betrayal creates a larger social disaster because the organizations and agencies in place to address the disasters respond to them in their routine ways, failing to gather information about the complexities of the social structure and thereby exacerbating the already existing problems. The routine response becomes overwhelmed, and further measures are taken to save an existing system that does not and will never meet the needs of the victims. Without a large-scale intervention to address and correct the structural problems that will help meet the needs of victims, many could die. One example of structural betrayal during Katrina was the failure of FEMA to adequately store emergency supplies outside of the disaster area to facilitate a quick and immediate response. Moreover, calls for qualified personnel were issued well after the disaster. By the time the existing response system was viewed as failing and reinforcements arrived, many officials were vested in the failing structure, creating feelings of betrayal on behalf of the citizens who were victims of the lack of structural response.

By August 31, 2005, the situation in New Orleans became more critical each moment. Lieutenant General Russel Honoré was designated the commander of the operations and was responsible for coordinating military relief efforts. By September, the sentiment of the relief efforts changed when Mayor Nagin said:

> Now, I will tell you this—and I give the president some credit on this—he sent one John Wayne dude down here that can get some stuff done, and his name is [Lieutenant] General Honoré. And he came off the doggone chopper, and he started cussing and people started moving. And he's getting some stuff done. (Transcript: New Orleans Mayor Ray C. Nagin's Interview, 2005)

In fact, from the moment Lieutenant General Honoré's command hit the ground, he was forced to reestablish order and create a fundamental plan in spite of the disarray caused by FEMA's inability to structurally manage the disaster response. Lieutenant General Honoré's actions went beyond the usual political recriminations and swift second-guessing that usually follow disasters. Most importantly, it took Lieutenant General Honoré one day to do what the entire Department of Homeland Security could not.

Equivocal betrayal is used by Gunter and Kroll-Smith (2007) when there is no clear evidence regarding the motives of government and corporate actions. It is difficult to understand this form of betrayal because it is rooted in the fabric of the community and in the political, social, and economic landscapes. Equivocal betrayal tends to be dismissed due to its lack of physical evidence, but it is real; although it is difficult to prove, it is the form of betrayal that remains within the community. We can trace premeditated betrayal to the person who is guilty of betrayal, which is oftentimes accompanied by a long list of deeds that include malfeasance, dereliction of duty, or reckless endangerment. Likewise, when structural betrayal is found, a clear chain of evidence can pinpoint the organizational failure that led to the secondary disaster associated with the failure of an adequate response. As stated earlier, with the dismissal of community members' claims, their situation becomes more dire as the "blame game" picks up momentum; the only potential for the community to have a voice is if the affected persons are able to address public bodies in hearings empanelled to investigate the previous two levels of betrayal.

After Katrina, several congressional hearings were held to uncover information regarding the actions of the government prior to, during, and after the flooding of the city. Dylan French Cole, or "Mamma D" as she is called by many of her 7th Ward neighbors, told members of the bi-partisan House Select Committee investigating the response to Hurricane Katrina, "I have witnesses that they bombed the walls of the levee, boom, boom!" (Stein & Preuss, 2006, p. 37). She was one of several people who addressed the committee. Others told the committee about the treatment they endured at the hands of those sent to rescue them, namely U.S. military personnel who kept the "refugees" behind barbed wire and refused to answer requests for first aid and used fear tactics to control people. One witness described the actions as follows: "they [the military] set us up so we would rebel, so that they could shoot at us . . . at one point they brought in two truck loads of dogs and let the dogs out" (Myers, 2005; Hodges, 2005; as cited in Stein & Preuss, 2006, p. 37).[19] Such testimony, coupled with numerous other oral histories accounting events shared by countless other survivors detailing the maltreatment of the evacuees, set the stage for equivocal betrayal. Although the motives of such actions, if they are true, are unknown, they lead to further speculation. Events such as the ones recorded on the *Alive in Truth: The New Orleans Disaster Oral History and Memory Project* website (*Alive in Truth*, 2005) echo the long-historical link of distrust that manifests itself. One *Alive in Truth* respondent stated the following:

> It didn't hit me until after I got here. I don't know, the Lord just gave me strength, just wanting to live. But a lot of prayers, a lot of faith and trust in God because, like I said, the military man, we had so many problems with the military. Going back to when I found my mom underneath the Causeway Overpass, I was told by the first person that came off the helicopter that they would transfer my mom, and I trusted this guy to do what he said, and he didn't do it. I lost trust in our military, I lost trust in our government officials, the leaders that said they were doing such and such. It wasn't being done, and a lot of people got separated because of that. My mom, she's safe now, but she left a lot of personal belongings, and I have to hear this every day. I have to watch her cry.

These sentiments echo the feelings of many of those in New Orleans, which is detrimental to the social vitality and future of the city.

Summary

After the disaster, the perceived lack of a sufficient response by all levels of government, and specifically President Bush, was viewed for weeks across the country via the mass media. The effects of Hurricane Katrina and the impact of the botched recovery efforts created images for entire generations of Americans who observed one of the greatest natural disasters in the history of the United States unfold. We will never be able to fully comprehend all the consequences of a loss of trust (be it a loss of trust in the physical landscape or a loss of trust in the political institutions that effect change in the political landscape) and how that loss of trust will effect us, but we postulate that increased civic trust leads to increased civic participation in various landscapes. Increased civic trust forms the basis for the place attachment to one's environment and how they react effectively. Scholars such as Brader (2006) and Marcus and Mackuen (1993) argued that our effective responses precede cognition when people make evaluations of events. In essence, people oftentimes feel before they think. Consequently, emotions can induce political leavening (Marcus, Neuman, & MacKuen, 2000), which increases political intent and participation (Brader, 2005), affects evaluation of the economy (Conover & Feldman, 1986), and increases the degree to which people feel positively or negatively about other people, civic leaders and government agencies and organizations that have failed or betrayed them.

The cycle of distrust becomes more insidious because the need to trust becomes replaced by the need for an increased militarized landscape, which is characterized by an overarching distrust by both the citizens toward the government and the government toward its citizens. To this extent, the civic breakdown in New Orleans is instructive because the post-Katrina environment in and around most severely flooded areas draws clear parallels to the current Iraqi war-zones as the fundamental organization and holding principle of society (Hardt & Negri, 2004). In both New Orleans and Baghdad, humanistic values have given way to militaristic values and the value, to some, seems to have diminished because death is evermore present and society increasingly organizes itself for the production of violence (Giroux, 2006; Mariscal, 2003). City neighborhoods, once flooded by water, become flooded with armed military national guardsmen and guardswomen with orders to shoot to kill.[20]

Notes

1. As cited in Gunter and Kroll-Smith, 2007, p. 70.

2. As cited in Cordasco, Eisenman, Glick, Golden, & Asch, 2007 p. 278.

3. "Furthermore, distrust in authorities among New Orleans' impoverished residents is rooted in local history. In 1927, the Great Mississippi River Flood was threatening to destroy New Orleans, including its critical downtown regional financial institutions. To avert the threat, and in part to stabilize the financial markets, it was decided to perform a controlled break in the New Orleans levees, thereby selectively flooding poor areas and saving financial institutions" (Barry, 1997, as cited in Cordasco, Eisenman, Glick, Golden, & Asch, 2007, p. 277). "This event lives on in the memories and oral history of the residents of the deliberately flooded areas" (Brinkley, 2006, as cited in Cordasco et al., 2007, p. 277).

4. One obvious example of this is seen in the forming of the Soviet Union and the election of Joseph Stalin to office, which led to a dynastic style line of appointment to leadership of the politburo. A more contemporary example is illustrated by Fidel Castro, whose rule will be carried on after his death through a manner of dynastic appointment to office.

5. See Post, 2006, p. 25

6. Historical experience in reference to African Americans and other minority groups was manifested in their subjugation and current-continual marginalization.

7. For more information see Avery (2006) and Welch et al. (2005).

8. According to Frymer et al. (2006), "Some of these institutions were designed to maintain racial hierarchies, and others were designed to avoid potentially divisive national conflicts over race. In either case, our governing institutions reflect the desires of the founders to avoid conflicts over civil rights (particularly for Blacks) and, thus, enable the maintenance of racial inequities" (p. 47).

9. See also Dahl, 1961.

10. Italics were added for emphasis.

11. The Vanport Flood of 1948 brought African American social inequities and issues of residential segregation to the forefront of national news. The Vanport Flood illustrated long-seated racial sentiments toward the well-being of African Americans in relation to Whites in that African Americans were situated in zones of sacrifice similar to the manner they were during the Mississippi Flood of 1927 and Hurricane Katrina (Rivera & Miller, 2007).

12. The Disaster Relief Act of 1950 allowed state governments to petition for aid in the event of a disaster; however, aid was not guaranteed to the petitioning state government. This opt-out option that the federal government has in reference to responding to a distressed state in the event of a disaster has been carried over into all disaster relief legislation since its inception in 1950 (see Rivera & Miller, 2007).

13. According to Barber (1983, as cited in Ritchie & Gill, 2007, p. 106), trust can be defined as "socially learned and socially confirmed expectations that people have of each other, of the organizations and institutions in which they live, and of the natural and moral social orders that set the fundamental understandings of their lives."

14. See also Bourdieu, 1980, 1983/1986; Burt, 1992; Coleman, 1988, 1990; Erickson, 1995, 1996; Flap, 1991, 1994; Lin, 1982, 1995; Portes, 1998; and Putnam, 1993, 1995.

15. Lin (2001) also noted that all of these factors combine to form control over a decision-making situation, which is why social capital is so successful.

16. Dixon et al. (2005) also argued that epistemological predispositions are "based on naturalist propositions, whereby social knowledge must be grounded in material phenomena and must take the form of either analytical statements derived from deductive logic or synthetic statements derived from inductive inference" (p. 6).

17. MacGillivary and Walker (2000) defined formal social capital as trust in organizations and social systems, specifically including governmental institutions and agencies.

18. Ritchie and Gill (2007) define a healthy community as containing high amounts of social capital, which is exhibited in trust, fellowship, associations, connections, networks, social intercourse, goodwill, sympathy, and norms of reciprocity.

19. See also Myers, 2005.

20. A fed-up Louisianan Governor, Blanco warned the lawbreakers that extra troops have already arrived in the city, and others are on the way—and they're "locked and loaded." She said Thursday night that 300 soldiers from the Arkansas National Guard had arrived—"fresh back from Iraq . . . These are some of the 40,000 extra troops that I have demanded," Blanco said. "They have M-16's, and they are locked and loaded . . . I have one message for these hoodlums: These troops know how to shoot and kill, and they are more than willing to do so if necessary, and I expect they will" (CNN.com, 2005).

Conclusion

The interplay among the physical, social, economic, and political landscapes of a geographical location impacts the development and evolution of each landscape and people's attachment to place. The physical landscape is the foundation for other landscape development because it presents the natural characteristics that society employs. The physical characteristics of a place initially define the interactional potential of the society living in that space, which allows specific interaction patterns to manifest in distinct ways that contribute to their interactional past. Societies develop a sense of place and a place attachment unique to individuals and their specific geographical location. Over time, the individual's interactional past, sense of place, and place attachment contribute to the society's overall interactional past. How society views its interactional past in a specific location influences its ability to alter the place's interactional potential through technology because it relates to subsequent redevelopment and rebuilding efforts.

The interactional potential of a place includes not only topological characteristics but also natural resources, such as rivers, forests, mountains, shorelines, and the weather. In the case of New Orleans, the presence of the waters from the Mississippi River and the Gulf of Mexico has enabled human societies to move throughout the western portion of the United States and gain access to international waters and commerce. These natural resources and the society's entrepreneurial spirit have led to the development of the place's social and economic landscapes. Furthermore, the natural interactional potential that relates to weather patterns, land fertility, and wildlife has directly influenced the area's social and cultural landscapes because society has adapted while contending with these conditions. By working with the natural interactional potential of the Mississippi Delta, New Orleans society has developed by incorporating and using resources specific to the area, such as seafood, agriculture, and river and coastal navigation, which has had profound cultural implications for the people who live in the area.

Furthermore, the natural interactional potential has directly impacted the society's interactional past. The availability of fertile land in the development of the region led to the institution of slavery, which has had profound effects on the current social and cultural landscapes of the city. Also, the area's potential to flood has had profound effects on the social and economic landscapes because, due to racial sentiments, lower income households have been located in areas of the city where there is a higher potential to flood, whereas higher income households have been located away from the flood zones. The potential for floods to occur has resulted in society altering the physical landscapes' natural topography, thus creating new interactional potentials. These new potentials coincided with residential and commercial development into areas of the physical landscape where human interaction was limited, but these potentials have also altered the social, economic, and political landscapes. New development has meant the expansion of the economy and enlargement of certain social classes, which has also influenced the dominance of one culture or group.

The interactional potential also directly affects a society's political landscape, which manifests itself through the geographic distribution of land by means of zoning and land usage codes; however, a place's interactional past has the most profound effect on the political landscape. The social experiences that occur within a society in a specific place, which are motivated by the physical, social, and economic landscapes, influence social actions and have political implications for the society. The social and economic landscapes' interactional past continually shapes the development of political landscapes that are inequitable, corrupt, and almost devoid of accountability.

New Orleans' interactional past, in relation to the natural physical landscape, has prompted the development of political landscapes that continually attempt to alter the place's interactional potential through technology. This technology includes finding ways to control flooding and minimize the effects of hurricanes to avoid devastating interactional pasts that may occur in the future. These alterations to the physical, social, and economic landscapes inevitably alter the political landscape because of changing interactional potentials and pasts that develop as a result of an alteration to the political landscape.

This course of logic may lead some to conclude that the political landscape is the paramount landscape that influences all others, but this is not the case. Although it is true that the political landscape can alter the interactional potentials and future pasts of the physical, social, and economic landscapes,[1] the interactional potential of the physical landscape is actually the most influential landscape. Attempts to address the potential of hurricanes and flooding can be made through the political landscape; however, from time to time, the physical landscape's potential to foster a catastrophic natural disaster, such as Katrina, is a testament to the physical landscape's influence on the other landscapes. Although all three landscapes continually interact and influence one another's development, the social, economic, and political landscapes are socially constructed and set in motion by humans.

The social, economic, and political forces temporarily exert control over the physical landscape to make it more conducive to human needs. The continual application of temporary alteration adheres to its natural interactional potential, which *cannot be controlled* by human society with any degree of certainty. One example of how society attempts to control the physical landscape is observed in the government's control over the flow and flooding of the Mississippi River, which has resulted in fewer flooding events; however, the events that have taken place have been far more destructive.

Even before the colonization of the area by European powers, the physical landscape dictated the development of cultures and societies through its interactional potential. During the city's history, the social, economic, and political landscapes that developed have continually attempted to augment the physical landscape while exploiting the local natural resources. By constructing levees and canals, the city attempted to mitigate disaster destruction, increase residential and commercial development, and stimulate the economy. Through the construction of the levees and canals, the augmented physical landscape brought new interactional potentials in the form of population and economic expansion, making the area more vulnerable to natural disasters.

The alteration of the physical landscape did not result in a parallel augmentation of its interactional potential. Because the levees and canals could not alter all of the interactional potential of the physical landscape, such as the occurrence of hurricanes and flooding, these events occurred anyway, and when they did, the population did not expect them. New Orleans illustrates what can happen when the initial interactional potential of a place is forgotten because the new technologically-enhanced physical landscape appears different. The case of New Orleans has taught us not to make bigger and better levees and continue to technologically enhance the physical landscape, but rather to work with the physical landscape and to be cognizant of its natural interactional potential.

The Postmodern View of the Environment

Human society must revert back to a more symbiotic view of the environment. A radical shift in perspective is needed so individuals and governments can understand that what is detrimental to the physical landscape is also detrimental to the social, economic, and political landscapes. To achieve an ecological view that enhances the value of the physical landscape, we must take steps to reeducate and rebuild our connection with the natural world as opposed to attempting to "conquer" it. Several researchers focus attention on the physical landscape's apparent absence from strategic management literature (Hosmer, 1994; Pauchant & Fortier, 1990; Throop, Starik, & Rands, 1993), the stakeholder theory (Starik, 1995), and in the field of business ethics (Gladwin, Kennelly, & Krause, 1995). Moreover, Gladwin et al. (1995) maintain that there is a lack of attention given to the physical landscape in management theory and education, which perpetuates notions of human existence outside of nature.[2] The belief that humans exist either outside of or above nature leads to the production and reproduction of human vulnerability in the urban landscape because it presupposes that humans are, for the most part, incapable of being significantly affected by natural hazards.

Gladwin et al. (1995) explained that most of the western world, and almost all the developed world, have adhered to the technocentric paradigm since the Scientific Revolution:

> Humankind is separate from and superior to nature. Humans are the only locus of intrinsic value. They have the right to master natural creation for human benefit. The objectified natural world thus has only instrumental and typically momentary quantifiable value as a commodity. Ethics are narrowly homocentric and utilitarian because contemporary and proximate human beings matter most. (Gladwin et al., 1995, p. 882)

Human society operates outside of the physical landscape and uses the environment for purely economic reasons. Current generations only need to pass on to the next an aggregate capital stock of resources that is no less than the one currently enjoyed (Gladwin et al., 1995). Human society views the physical landscape's resources as inexhaustible within this perspective. This is not to say there is an infinite amount of oil, gold, or other manner of natural resource but rather that human ingenuity always finds a way to exploit nature in an effort to find substitutes for diminishing resources, which implies that resources are limitless. Moreover, the overall growth in population worldwide is not seen as a cause for environmental degradation but rather as positive and needed for the development of creative solutions to compensate for specific diminishing natural resources (Simon, 1981).

From an economic standpoint, and adhering to the technocentric paradigm, markets operate in a closed-linear system, isolated from nature and where exchange value occurs between industries and households—everything else is exogenous (Gladwin et al., 1995). Cities in this framework are seen as engines of economic growth in which any mention of sustainable development is used to justify trends already well established in urban development (Pelling, 2003; World Bank, 2000). This notion supports the expansion of worldwide human population.

> The world is largely empty. Growth is good, and more growth is better; growth enables governments to tax and raise resources for environmental protection and leads to less polluting industries and adoption of cleaner technologies. (Gladwin et al., 1995, p. 884)

Although the logic behind economic expansion through population expansion is theoretically sound, some scholars believe that economic perspectives based on the technocentric paradigm perpetuate poverty and underdevelopment, the growth of economic and social disparities, the bestowment of privileges to a wealthy minority at the expense of the human majority, the exhaustion and dispersion of a one-time inheritance of natural capital, the reduction of the rights of future generations, the legitimization of the concentration of economic and political power, and the separation of the control of productive assets from the communities that depend on them (Daly & Cobb, 1994; Ehrlich, 1994; Gladwin et al., 1995; Korten, 1990; Maclean, 1990; Sen, 1982; Weiss, 1989). This is not to say that there has not been recent acknowledgement by economists and government officials worldwide that there *is not* an inexhaustible amount of natural resources or capital; however, the acknowledgement has done more for the research and development of finding alternative natural resources to exploit when the currently used resources are gone as opposed to attempting to live more symbiotically with the physical landscape.

In response to the technocentric paradigm and its alternative environmentalist paradigm, Gladwin et al. (1995) proposed a somewhat integrated paradigm of the two extremes. According to the sustaincentric paradigm:

> Humans are neither totally disengaged from nor totally immersed in the rest of nature. Although they are part of the biosphere in organic and ecological terms, humans are above the biosphere in intellectual terms. . . The crucial conse-

quence is that humans "have become, by the power of a glorious evolutionary accident called intelligence, the stewards of life's continuity on earth. We did not ask for this role, but we cannot abjure it. We may not be suited for it, but here we are." (Calvin, 1994, as cited in Gladwin et al., 1995, pp. 890-891)

From this perspective, humans have been entrusted with the task of managing the environment over the course of human existence as opposed to excessively exploiting it for contemporary needs. It is understood within sustaincentric paradigm that the physical landscape and all of its resources are finite and vulnerable to human interference. Therefore, the human population must be stabilized and the consumption of products in developed countries must be scaled down to sustain the integrity of both natural and social life-support systems through logistic growth (Holling, 1994).

As opposed to the technocentric paradigm, the sustaincentric paradigm maintains that the economy is underpinned by the physical landscape and that changes to one inherently affect the other. Ecological and social externalities are internalized by markets that require the efficient allocation of resources, putting into place other policy instruments and economic incentives that limit pure market situations and place strain on natural resources and basic human satisfaction (Gladwin et al., 1995). Moreover, material and energy growth are bound by ecological and entropic limits, and they cannot continue in a closed system.[3] The reason for this is that the benefits of past growth have not been fully realized by human society because the richest 20 percent of the world's population possess 83 percent of the financial wealth and consumes 80 percent of the world's resources (United Nations Development Programme, 1994). For humanity to fully satisfy its basic needs, it must first distribute what it has gained through its past development.[4]

The concentration of the benefits of past growth in the hands of a few has perpetuated society's *need* to further exploit the physical landscape in the contemporary world. It is this lack of equitable distribution of existing resources that causes human society to develop in geographic areas that are prone to natural hazards. The human development of physical landscapes considered high risk areas is not fundamentally due to an affinity to live in these areas; it is because there is a *perceived* need to develop in these areas because preexisting development has not yielded enough benefits for the majority of society.

Altering Society's Environmental Perspective and Promoting Social Change

Despite the low probability of a redistribution of wealth and resources for society's benefit, it *is* possible, to limit the exposure of environmental threats. To this end, humanity should employ a perspective that works with nature and acknowledges the preeminence of the physical landscape in development. If humans subject themselves to the hazards of nature by electing to develop in high risk areas, they must be cognizant of how to live in harmony with the physical landscape in a way that allows for place-based development.

CONCLUSION

> Construction interferes with the land-building process: levees contain the silt needed to replenish the lowlands, dredging loosens the land by killing freshwater plants, floodgates and reservoirs further aggravate marsh subsistence. "It's ironic. . . . The system which brings prosperity and security to humans is literally costing them the earth beneath their feet." (Brown & Perkins, 1992, as cited in Shallat, 2006, p. 103)

By allowing the physical landscape to naturally evolve through time, natural mitigation against hazards will manifest themselves; however, when humans attempt to control the contour of the land by shaping it solely in humanity's interest, natural mitigation will not occur, which leaves a region defenseless against mega-disasters.[5] Human-built controls further promote destruction caused by natural disasters.[6]

A society's ability to withstand major damage and disruption from a predictable characteristic of the physical landscape illustrates the fact that the society was developed in a sustainable way (Oliver-Smith, 1996).[7]

> natural disasters are signifiers of the inequalities that underpin capitalist (and alternative) development of unsound and manifestly unsustainable human–environment relations. . . . [U]rban disasters are not amendable to technological quick fixes alone, and rather that the nature of disaster risk is constantly being redefined as changes to urban landscapes and socio-economic characteristics unfold. Urbanization affects disasters just as profoundly as disasters affect urbanizations. (Mitchell, 1999, as cited in Pelling, 2003, pp. 6-7)

For a society to develop in a sustainable way, it must view disasters and hazards as integral parts of environmental and human systems (Oliver-Smith, 1996); however, as discussed above, developed human society has moved away from viewing the environment in this way and views disasters and hazards as random phenomena outside of the normal environmental system. According to Pelling (2003, pp. 7-8), human movement away from nature combined with the local effects of current global environmental change and in addition to economic, political, and cultural globalization adds to greater uncertainty in development planning, specifically in reference to the prediction and management of natural hazards and human vulnerability. Alternatively, in more traditional societies, disaster is viewed as a fundamental characteristic of the environment. People living in these societies develop reasonably effective adaptation strategies that allow them to maintain long-term and viable ways of life in difficult and sometimes inhospitable conditions (Torry, 1979).

Traditional societies are capable of maintaining themselves in relatively difficult conditions because they seem to exhibit greater local understanding of their social and physical environments than societies that are considered more developed. This allows traditional societies the opportunity to reduce both short-term and long-term losses in the event of a disaster (Gordon, Farberow, & Maida, 1996; Guillette, 1991; Loughlin, 1995; McSpadden, 1991; Oliver-Smith, 1986, 1996). This is alternatively illustrated by more developed experiences with reoccurring hazards in which the population suffers due to either a refusal or simply an inclination to disregard local understanding of the physical landscape.

> Mitchell (1999) found this to be the case for several of the world's largest cities including Seoul, Republic of Korea, and Dhaka, Bangladesh, where despite hundreds of lives having been lost to floods neither city had collated information on hazard experience as a first step to mitigating risk. (Pelling, 2003, p. 14)

Understanding of the physical landscape decreases the number of displaced residents and shifts the effects of a disaster onto alternative physical landscapes that are more capable of enduring them.[8]

Local understanding of their physical landscape is paramount if societies and cities are to become less vulnerable; however, the general public must alter their perception of responsibility for this to come to fruition. According to Pelling (2003), citizens tend to view environmental problems and their vulnerability to natural hazards as the responsibility of the government, and they therefore feel that disaster preparedness should not be a priority in their daily lives. By adhering to this idea, citizens perceive that they are forcing responsibility for mitigation and the limiting of vulnerability to higher echelons of the national governmental system.

However, this perception is detrimental to their existence because, in United States' practice, the federal government views the responsibility of mitigation and limitation of vulnerability as the specific responsibility of the local citizenry. The federal government's perspective is based on the notion that localities have the best understanding of their physical and social landscapes, which logically dictates that they are best suited to plan and develop in a way that limits vulnerability to the general public. Therefore, the notion that the federal government was responsible for the shortcomings of the mitigation efforts of New Orleans is flawed because the system is structured so that the leadership of New Orleans is directly responsible.

The placement of responsibility for local issues of disaster vulnerability on the federal government is a manifestation of the local citizenry's movement away from understanding and reciprocal interaction with the physical landscape; therefore, they feel as though someone else has a better understanding or knowledge of how to deal with phenomena generated in the physical landscape than they do. Lack of knowledge of the physical landscape is a result of the public's disassociation with the physical landscape from their daily lives. This disassociation contributes to the public's opinion that concern for the landscape should not occupy their daily lives.[9] For citizens to accept responsibility for their physical landscape, they must return to a situation where they continually interact with it and derive meaning from it daily.[10]

Life in an Age of Mega-Disasters

Although it is important for society to understand its social environment so disaster response can be better coordinated, it is more important for society to understand the physical landscape. Such an understanding leads to a more accurate prediction of disasters and more effective responses to events, such as evacuation or coordination of search and rescue efforts. An ecologic–symbolic approach to understanding landscape change helps individuals in a society become more adept at living with the occurrence of natural disasters in an age of more inevitable and complex mega disasters. Understanding of the physical landscape comes from practical activities in the environment (Butz & Eyles, 1997; Ingold, 1992). According to Gibson (1979), direct perception and environmental understanding come from the notions of environmental affordances.

> "[The environment] *offers* the animal what it *provides* or *furnishes*, either for good or ill" (Gibson, 1979, p. 127; emphasis in original) for the consummation of behavior. These affordances exist as inherent potentials of environmental objects themselves, independent of whether or how a subject uses them. Thus, environmental objects are not neutral objects waiting for individuals to assign them meaning; their meaning is what they afford. (Butz & Eyles, p. 7)

By understanding that the physical landscape offers certain affordances, either through natural resources or weather patterns, and that these affordances are inherent and unchanging characteristics of the landscape, urban development can evolve in a way that takes the affordances into account. By considering this during the process of urban development, occurrences that are destructive to both the landscape and the built environment can be avoided.

In essence, "development" means enhancing what is already present, which can also revitalize and reinvigorate places. Moreover, development can initiate processes of economic, social, and environmental sustainability, but most times it does not (Day, 2002). In most cases, development separates humans from the inherent natural aspects of the physical landscape (Day, 2002). The Mississippi River serves as a mental and physical landmark that shaped the ideas about the city and its spatial characteristics (Kelman, 2003). However, in the pursuit of economic development, New Orleans' citizens have attempted to control their physical landscape over time to make the unpredictable predictable. The pursuit of economic vitality resulted in the physical and mental distancing of the people from the Mississippi.

> Whereas the Mississippi formed a part of the daily lives of most New Orleanians from the colonial period until the era of Reconstruction, for a century after that the river and its waterfront were off-limits or inaccessible to most of the city's residents. The artificial levees and warehouses lining the waterfront formed a physical barrier between New Orleans and the Mississippi, while the conflation of the riverfront's public character with commercial endeavors further alienated many people from that space. In short, by the mid-twentieth century, sepia-toned images of the river flowing by the city were little more than vestiges of public memory lingering from the antebellum era. (Kelman, 2003, p. 15)

This physical distancing has led to feelings of security among residents, which are occasionally challenged by random flooding and their limited knowledge of contemporary vulnerability to natural disasters.

Day (2002) maintained that a location's sense of place develops slowly over time and is always changing, but when the location is obliterated by an event, reestablishing a sense of place takes a long time—sometimes several generations.[11] Therefore, reconstruction of the city needs to occur to return to a sense of normalcy in a timelier manner while reestablishing New Orleans' sense of place along sustainable lines. City planning must be designed along these lines so that it allows all residents to interact with the physical landscape on a daily basis. The social relationships people have with the forces of nature and the meaning people ascribe to their interactions produce the everyday experience of place for any locale (Burley, Jenkins, & Azcona, 2006; Sack, 1992; Williams & Patterson, 1996).

By interacting with the physical landscape, both new and returning residents of New Orleans can develop a sense of place that is more conscious of the natural location as opposed to the manmade landscape that will be constructed. Residents should be able to interact with the physical landscape. Learning from the physical landscape will decrease residents' vulnerability to hazards simply because they will be more aware of the actual risks. Residents will not have to depend on government officials to tell them there are impending risks to their safety, which will allow them to make more informed decisions about confronting disaster situations. This is not to say that residents were not cognizant of the occurrence of disasters in New Orleans, but the knowledge of severity was not understood. What has occurred in the aftermath of Katrina is an increased acknowledgement that residential displacement is ominously possible with the potential for future storms; this acknowledgement was not present to the same degree in the past (Burley et al., 2006).

Reestablishing a sense of place—and, subsequently, place attachment—among both returning and new residents is paramount to the city's future success. The problem becomes how to establish place attachment to a location where residents have acquired a heightened sense of vulnerability that is perpetuated by feelings that residents have "no voice" in the city's restoration (Burley et al., 2006):

> Within the particular problem that southeastern Louisiana is facing, now more so than ever, they are understandably untrustworthy of outsiders who they feel make little attempt at including them in restorative processes. If practitioners, in restorative processes in general, want residents "on board" for restoration projects and policies, then they must actively involve residents in actual restoration. (p. 38)

By incorporating residents in the actual implementation processes of reconstruction and restoration projects, they will develop a heightened sense of attachment and respect for their neighborhoods and city (Hutter, Miller, & Rivera, 2006). Moreover, the inclusion of residents in reconstruction and restoration gives them some potential for participation in decision making, which fosters a sense of ownership about the project (Hutter et al., 2006). As advantageous as it might be to develop place attachment and the local economy by having city residents participate in reconstruction and restoration, bringing residents into policy and decision-making processes would further enhance place attachment by instilling a sense of direct ownership over government functions.

Heightening a sense of direct ownership of local government can be achieved through a process called *reflexive inclusion* (Miller & Rivera, 2006a).

> reflexive inclusiveness refers to using knowledge in the development of sensitivities for all aspects of modern life, particularly characterized by the ongoing

problems that exist surrounding the rebuilding and repopulation of New Orleans. A reflexive inclusion actively involves the citizenry by educating the public, empowering them to give voice to issues, and places them at the center of the decision making process by establishing a symmetrical understanding of the negative public perception. (Lhulier & Miller, 2006, as cited in Miller & Rivera, 2006a, p. 43)[12]

Reflexive inclusion can only work toward increasing place attachment and civic trust in the sense that the process is fully implemented and not attempted to temporarily appease public sentiment. To this end, piloting a change in government procedures of this kind would need to be long-term to fully evaluate its effectiveness at addressing local policy issues and its actual degree of acceptance on the part of government bureaucrats and the public. Moreover, the success of reflexive inclusion hinges on adhering to three guiding principles: transparency, sustainable equity, and the maintenance of a results-based culture (Miller & Rivera, 2006a). By adhering to these principles, local governments become more accountable of their actions and decisions while at the same time enhancing civic trust among its population. Investing in a process such as reflexive inclusion can aid residents in acquiring a sense of self-determination for their futures, directly molding their degree of vulnerability and the city's course in the future.

Some Final Thoughts

The previous chapters combine various theories ranging from the social construction of reality and how humans cognitively organize their lifeworlds to the political organization of land use among stakeholders. For practitioners to be more effective in developing a response to catastrophic environmental change using a layered landscape framework, we must first recognize the fact that different landscapes have a variety of meanings to different people and groups. Also, these landscapes are symbolic reflections of how we as individuals define our immediate surroundings, while the larger landscapes or cityscapes serve as complex arrays of cultural, political, economic, and social expressions. Disasters such as Hurricane Katrina require a renegotiation of the socially constructed complex expressions; it is these renegotiations that alter the social structure and bring about social change. We maintain that place attachment to the physical environment facilitates an understanding of changes taking place as the redevelopment occurs in the midst of renegotiating the factors that govern life after a disaster and in light of the ongoing social, cultural, political, and economic changes.

Katrina is a triggering event not only because of its occurrence as a hurricane, of which there are many each year, but also because of its ability to call the nation's attention to the plight of a region. The environmental devastation and Katrina's intensity were a wakeup call. The global occurrence of historically unprecedented natural events can signal worse environmental disasters in the future. If stakeholders fail to learn from the intensity of Katrina and use place-based knowledge to aid in rebuilding a connection to place, they will continue to suffer from more frequent destructive natural occurrences that damage the social fabric. In the future, environmental assaults and their ability to disrupt life with increasing frequency and intensity will present unprecedented challenges to survivors' ability to integrate place-based knowledge and cope with the challenges of recovery and redevelopment.

Place-Based Knowledge

Throughout this book, we have argued that places have the ability to instruct societies on how to best live in a specific location; that the physical landscape dictates unique interactional potentials, which in turn elicit unique interactional pasts. The anthropocentric views that have led to the economic conquest of nature have also led to continued human development in high-risk areas, exposing an increasing number of communities to extreme natural risks. Modern society's ability to cope with extreme natural phenomena, due to a continued disassociation with the physical landscape through the use of technology, is at the root of increased vulnerability to catastrophic events. Moreover, we have also mentioned the relatively high degree of coping skills that traditional societies, in comparison to modern society, have used in regards to natural disasters. This is due to their ability to use placed-based knowledge as opposed to technology.

Figure 7.1 illustrates the ways in which the landscapes we have discussed here interact with one another within traditional societies. As depicted, the landscapes we discuss are layered atop the physical landscape because it is what molds the development of all the others. Place attachment develops as a direct result of the physical landscape and is, therefore, depicted as layered directly above the physical landscape. However, the ideology of the traditional society cuts across all landscapes, which is rooted in culture and historical experience. This ideology is based on a subsistence economy, which interacts with place attachment and, in turn, influences the development of the socially constructed landscapes and the perception of the physical landscape. Because ideology is influenced by place attachment in addition to culture, and more specifically historical experience within the physical environment, traditional societies use their place-based knowledge to cope with natural disasters and to develop their political, social, cultural, and economic landscapes.

In more developed and modern societies, a gap has been created between the social and cultural landscapes and the physical landscape (Figure 7.2). This gap has been caused by the use of technology to augment and conquer the physical environment and has shifted the overall influence of the physical environment to societies' social and cultural landscapes. In contemporary developed societies, place attachment is more closely influenced by the social and cultural landscapes than it is by the physical landscape. Moreover, the ideology that cuts across all of the landscapes is rooted in economics as opposed to culture and historical experience, which has led to political and economic landscapes that are also less influenced by the physical landscape, as illustrated by the gap between the physical landscape and all of the others. Place attachment and ideology influence each other's development; however, because ideology is strongly influenced by economics as opposed to the physical landscape, developed, modern societies view the physical environment as something to be exploited for the

The Traditional View

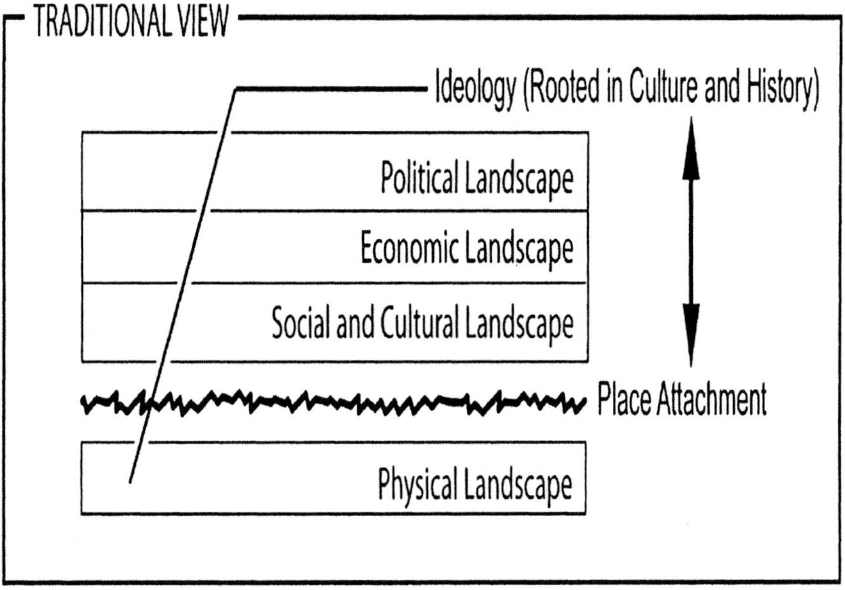

TRADITIONAL VIEW

- Ideology (Rooted in Culture and History)
- Political Landscape
- Economic Landscape
- Social and Cultural Landscape
- Place Attachment
- Physical Landscape

Figure 7.1. Graphic courtesy of Karlton Hughes.

benefit of the other socially constructed landscapes, as opposed to using its "knowledge." Therefore, economics and technology drive decision making with little or no respect for the natural physical landscape and its interactional potentials and pasts. This format places society in jeopardy of being destroyed by the physical landscape because they do not use place-based knowledge.

We are proposing that future societies must close the gaps among place attachment, the social and cultural landscapes, and the physical landscapes and return to an ideological model that uses place-based knowledge (Figure 7.3). In this model of society, the gaps between the physical landscape and all others is decreased by the use of sustainable development practices, which by definition denotes an increased knowledge of the influences of the physical landscape to guide the development of all of the other landscapes. In this future scenario, place attachment, as well as all other landscapes, will be more closely influenced by the physical landscape than they are in contemporary modern societies, which are depicted by their layered proximity to the physical landscape.

Society's ideology that cuts across all of the landscapes is rooted in place attachment, historical experience, and culture and dictates the use of more placed-based knowledge in relation to contemporary society. Although the optimal scenario would be the use of placed-based knowledge similar to traditional societies (as expressed by negligible gaps between the physical landscape and the others), societies will never be able to return to such a situation because they have developed beyond subsistence-based cultures that directly rely on the physical landscape.

The Modern View

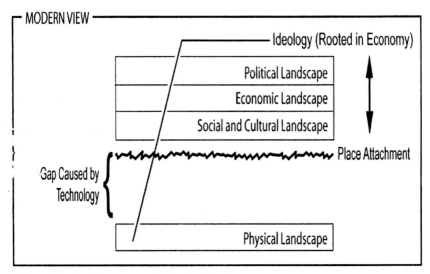

Figure 7.2. Graphic courtesy of Karlton Hughes.

Therefore, contemporary society's current ideology, which is rooted in economics, must again be augmented so it becomes rooted in place attachment, historical experience, and culture. To accomplish this, society must change its perception of the physical landscape, which is only possible from the bottom up due to the top's investment in the status quo. For this to happen, people must have a larger commitment to the development of their socially constructed landscapes by being more accountable for the governance and decision-making practices through a process similar to reflexive inclusion.

Radical Social Change

Adhering to a process similar to reflexive inclusion currently requires a sense of ambivalence on the behalf of agencies, local governments, and Congress because it is not legally necessary to directly include the public in policymaking. According to the Administrative Procedure Act of 1946, the public is allowed to petition agencies to craft a "rule" or policy. However, the act did not instruct an agency on how to handle or even address such petitions (Kerwin, 1999). In addition, it did not motivate agencies to expand their interactions with the public or limit them; the legislation left public interactions towards policymaking to the discretion of individual agencies. Furthermore, Kerwin (1999) maintained that the act was consistent with the political American ideology of the time that placed significant faith in the capacity of the government to solve public problems. Since the enactment of the Administrative Procedure Act, there has been significant progress in the incorporation of the public in the policymaking process; however, this progress has come with explicit limitations on public

The Future View

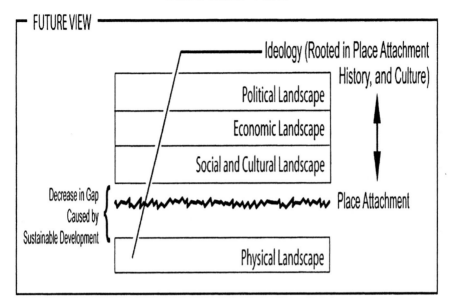

Figure 7.3. Graphic courtesy of Karlton Hughes.

participation regarding some key issues and under certain conditions, such as emergency situations.

Under the current system of policymaking, the ability of the public to participate in decision making or policy creation is dependent on the amount and intensity of public interest that the proposed policy is likely to generate.

> In instances of complex of highly controversial rules, the public hearing may be selected because it allows agency personnel to go to the field and explain what they are doing to affected parties and make a case for it. At other times the opposition may be so intractable, and predictable, that public meetings would serve little purpose other than catharsis. (Kerwin, 1999, pp. 83-84)

This government stance implicitly acknowledges that the public should be brought into decision-making discussions only when there is possible negative political backlash, such as negative media coverage or shifts in voting patterns. Moreover, this idea predicates that the public will be significantly aware of the policy and be directly affected by it in their daily lives to the degree that they are emotionally motivated to voice an opinion. Therefore, in situations where either agencies believe that there will be negligible political backlash or the public does not feel they will be explicitly affected by the policy, public participation is not typically sought by the government or by other policy-crafting agencies. In specific reference to local disaster mitigation and relief policies, public participation in the crafting process may need to be mandatory because the policies, although they may not directly affect the entire public's day-to-day activities, will severely affect the public under certain circumstances. The impact of these policies on the public is profound and inherently affects the ability of the public to address life and death situations, which further justifies their inclusion. In Louisiana, where non-structural mitigation laws and regulations are not incorporated in the state's statues and home rule laws—giving local governments the power to enact zoning laws, building codes, and subdivision regulations in flood-prone areas as they see fit—(La. Rev. Stat. 38.84., as cited in Mittler, 1988), public participation is needed to influence local authorities to enforce regulations and laws so they are not merely symbolic but actually limit damage caused by disasters.

Public participation in policy development is needed not only for disaster mitigation and relief but also in all types of public policies that directly affect the public; however, participation places a higher degree of accountability on the public as opposed to the government. Therefore, mandatory public participation in policy development, beyond that of a circulated petition or letter writing campaign, would be beneficial for the public's acceptance of policies while taking some weight off government accountability in the event of ill-constructed policies. Mandating public participation outside petition and letter writing would obviously affect bureaucratic processes, especially in reference to the amount of time spent on the formation of a policy; however, the time spent would gain wider public acceptance and hopefully save lives. Mandating local governments and agencies to include a certain degree of public participation in developing policies would also help the broader society.

Inherent in the notion of public participation is the idea that the public *will* participate given the opportunity. Public participation in their decision-making and policy-crafting processes relies heavily on the public's *trust* and investment in the process for the advantages of their inclusion to be realized. Minor alterations of the process to encourage participation by the public would be a largely symbolic gesture if the public declined to take advantage of the opportunity. Therefore, in accordance with making public participation mandatory on the behalf of local governments and agencies, the government should make voting and other forms of civic engagement compulsory to yield the most benefits from the practices of democracy and ensure that citizens' voices are heard.

An educated and civically-engaged citizenry can reduce their vulnerability to disasters. Citizen involvement can influence change in the way individuals, communities, states, and the federal government perceive human vulnerability to disasters. Through this bottom-up approach, radical social change in the political and social landscapes (i.e., a change in the perception of disaster mitigation and social vulnerability) can decrease and ultimately close the gap between bureaucratic and emergent norms.[13] Although there may be some drawbacks to financing and implementation, a bottoms-up strategy is the most practical course of action because of the amount of investment the higher echelons of society already have in current development strategies and environmental exploitation (Anderson, 2006).[14]

Change will never come by leaving it to the political elite to enact radical social change because they are beneficiaries of the current situation. It is only when the public participates in crafting policies through civic participation and voting that the institution of government can be modified in a way that specifically benefits the population in the manner that they wish. Without this participation, the government will continue to govern from an elitist standpoint in which they believe they know what the people want and what is in their best interest. From historical experience, what those in power believe the majority of the population needs and wants can sometimes be detrimental to the whole of society, and sometimes it could not be any farther from the truth (from the perspective of the masses).

The ever-changing political landscape may be the most difficult to understand or adjust to in the wake of the handling of the Katrina catastrophe. With the political landscape in such disarray, many citizens were left asking: Are local and state governments hopelessly inadequate to deal with disasters? The politics of disaster rebuilding in the political landscape began to form in the national consciousness as early as the 1880s, when a natural disaster led to a breakdown of congressional retreat; in 1886 and 1887, Congress appropriated $10,000 for the distribution of seeds to farmers after a massive drought (Olasky, 2007). In response to this, President Grover Cleveland vetoed the bill by asserting that it was wrong for federal funds to be used for such "charity," expressing "a hope that private and church philanthropy would come through" (Olasky, 2007, p. 33). Private sources of assistance in the form of relief contributions in the United States came from charitable sources. Hence, the debates raged in the 1800s and 1900s about the degree of federal and local involvement and are now continuing into the 21st century.

One thing that remains true about the response to natural disasters is the reality that governments at the local, state, and national levels tend toward the reactionary and, at best, act in a fragmented and costly manner. More specifically, the United States has taken steps after the World Trade Center attack on September 11, 2001, to create a more complex, federal disaster apparatus that includes rules layered with strategies, policies, and laws. According to Platt (2000), these strategies are intended to operate in conjunction with state and local governments as well as nongovernmental organizations and the private sector as a "partnership." However, some of these programs, while well intended, oftentimes undermine one another. For example, in some parts of state government, there is a push to restore the Louisiana wetlands, whereas parts of the federal government are seeking to dredge and channel the Mississippi River, thus adding to the coastal wetlands depletion problem. Also, the political landscape is often tempered and is the environment for many conflicts over the nature, character, and pace of the rebuilding efforts.

The cultural landscape has a profound effect on the region's development because it is intricately linked to the regions' economy. The perception of hurricane threat and imminent danger along the Gulf Coast remains real, and it has real consequences for the nation. If one perceives the threat of disaster as real and the factors associated with evacuation as available, such as personal finances and a place to evacuate to (i.e., extended family availability of hotel rooms), then the likelihood of evacuation is high. However, when there is a lack of a perception of threat and a lack of factors associated with evacuation, then the likelihood of evacuation is low.

The term *disaster subculture*, introduced in the 1960s and further developed in the 1970s, refers to a cultural adaptation in coping with recurrent threats and to the cultural defense used by a group to adapt to cognitive, behavioral, individual, and collective behaviors used by people in response to a disaster that has struck or has the potential to strike in the future (Moore, 1964). Later, researchers Werger and Weller (1973) cited three factors that facilitate the emergence of such a subculture: (1) the community must experience repeated disaster impacts and perceive them as repetitive and chronic; (2) there must be a period of forewarning; and (3) the damage must be perceived as widespread and is able to affect any part of the community. One time change can result from increased civic participation in the political landscape.

Over three decades ago, Hannigan and Kuenemar (1978) found that as the government accepted greater responsibility for flood mitigation, individual interest in flood-related matters weakened. A weakened response to perceived threats or a citizenry lulled into a false sense of security becomes less prepared to adapt because organizational responses appear to be sufficient. Nigg and Mileti (2002) asserted that underestimation of the threat of natural disasters contributes to an under appreciation of the threat, leading to a "sense that [the] government is responsible for reducing natural hazards and risks" (p. 283).

As hurricane seasons passed and large storms made landfall (leaving the region with moderate to miniscule devastation or totally unscathed), the sense of trust in the land to guard against flooding or technology to keep the waters at bay took hold. Natural protective barriers and manmade levee systems reinforced a view resulting in a false sense of security. Many residents remember that although some flooding and loss of life did occur during Hurricanes Betsy and Camille they still survived. Not only did they survive, but they were able to rebuild after the two most deadly natural disasters of their time.

However, many do not take into consideration the radical landscape changes taking place at such an alarming rate with both the Mississippi River and the erosion of the barrier islands and coastline. The factors that contribute to a disaster subculture are clear and present. The levees provide a barrier that is visible and has a place in popular culture; however, coastal erosion and the channeling of the Mississippi River, while not a part of the daily experience of the storms, are also not a part of the calculation of personal vulnerability. The natural barrier, which provided protection against certain disasters, along with a host of factors such as climate change (namely global warming), while not calculated as residents decide on their plans to evacuate, lie at the center of vulnerable societies built in a fragile environment.

The disaster subculture is important because it serves as a foundation for how to interact with nature and whether or when to evacuate during times of calamity. Some people chose not to evacuate based on a variety of reasons and they elected to remain home in spite of warnings. We maintain that unless a change in the disaster subculture occurs, survivors will not take their level of vulnerability to natural disasters seriously. Katrina brought into view the level of vulnerability that occurs when societies are ecologically disconnected and overly dependent on technology, while exposing the disaster subculture's effect on one's chances of survival when weathering the storm. A disaster-lax subculture combined with economic hardship and a lack of social support led to many deaths.

From a social–scientific perspective, Katrina created a host of problems located in a certain space that impacted the attachments people experience with the places in which they live. Detachment from the landscape is paramount to the future of the region. We maintain that there will be a loss of a unique cultural heritage within the American cultural fabric without a connection to the land. When detachment from the places that define the pre-disaster experience are widespread and pervasive, an environment for the practice of diversion (detournement) develops as business interests and political momentum builds. We have argued throughout this book that, in the wake of Katrina, whole neighborhoods have become vacant in the physical and conceptual sense and fall susceptible to being diverted or reappropriated, thereby putting them to different uses than their original marsh and swamp land environment. This reappropiation will occur when a space is not well integrated into the social, political, cultural, and physical landscapes of the city.

Often, when leaders fail to recognize the legitimacy or "standing" of such spaces and fail to equally affirm the rights of those in some parts of the community over others, the space becomes a capitalist commodity and economic entity, dominated by the interests of the bourgeoisie and politically elite rather than by the interests of those who were most closely associated with the land. Citizens become alienated, disconnected, and estranged from civic engagement in "islands of inopportunity." This form of spatial alienation radically limits the citizens' connection with the region's density and the city's social, cultural, political, physical, and economic landscapes and robs them of that which makes New Orleans and the Gulf Coast region unique. Key to understanding the reconnection of citizens is a deep appreciation of the historical development of landscapes and the maintenance of place attachment, even in the face of such profound devastation, as caused by Hurricane Katrina.

Notes

1. Alterations of interactional potentials of a physical landscape, and subsequent future interactional pasts, are achieved politically through the passing of legislation that allows for the augmentation of the physical landscape.

2. See also Buchholz (1993), Orr (1994), Stead and Stead (1992).

3. Hewitt (1983) maintained that natural disasters should be viewed as part of an ongoing relationship between society and nature, not as set-off, and that extreme disasters do not take place outside of development. In this way, the macro-economy and national political regimes/agendas contribute to the production of disasters (Pelling, 2003).

4. Because it is improbable that the wealthy will benevolently distribute their resources to the rest of society, the World Bank (1990, as cited in Gladwin et al., 1995) proposed a shift in the predominance of certain economic sectors and the use of taxes and other public policies:

Taxation and other public policies are shifted to favor labor intensity over capital intensity and to promote income and saving versus energy/matter throughout. Poverty reduction in sustaincentrism depends on "two equally important elements." The first element is to promote the productive use of the poor's most abundant asset—labor. It calls for policies that harness market incentives, social and political institutions, infrastructure, and technology to that end. The second is to provide basic social services to the poor. Primary health care, family planning, nutrition, and primary education are especially important. The two elements are mutually reinforcing; one without the other is not sufficient. (p. 893)

5. Pelling (2003) maintains that linkages between urbanization and disasters are weakly theorized, and as a consequence, mitigation is rarely integrated into urban development policy. However, as seen in New Orleans, when it is integrated into policy, the policies are not necessarily integrated to protect the already vulnerable segments of a society or city, but to create new development that places new populations in vulnerable situations. In this way, the integration is not mitigation but an attempt to control the physical landscape so that additional development can take place, which does not particularly correlate to less vulnerable populations and geographic areas.

6. "Of course, the tragedy of New Orleans is not exclusively, or even primarily, a technological one. Business and politicians did more than acquiesce in decisions to build in unbuildable areas. A political structure fraught with mendacity and a class system tinged with racial antipathy contributed to the magnitude of disaster. Hurricane Katrina occasioned a classic complex-systems failure. All the systems necessary to survive and recover from it—the drainage pumps, the first responders, the communications systems, the power grid—ultimately failed, one by one, a pitiful situation aggravated by the late and inadequate response of the national government. But the key event was a technological failure: the collapse of the canal walls" (Kolb, 2006, p. 110).

7. See also Castree and Braun (2001) and Hewitt (1983).

8. Pelling (2003) suggested that when risks are locally mitigated, the disaster is shifted on other populations and ecological systems, which simply shifts the vulnerability of one population and one ecosystem onto another. What we are arguing is that true understanding of the physical landscape would enable the society of a city to shift the effects of a disaster, through mitigation, to an area that is more capable of dispelling the hazard's destructive effects without making other populations or geographic locations more vulnerable.

9. Berke (1998) argued that it is not surprising there is a shortage of mitigation programs given the variety of obstacles due to a lack of interest in natural hazard risks.

10. Kolb (2006) noted the acknowledgement of the physical landscape in eighteenth-century Louisianian architecture:

A typical eighteenth-century Louisiana house (say, Madame John's Legacy, on Dumaine Street) attests to the realities of living there. The first flood is intended for storage only, the living area on the second floor is raised far above the ground, the walls are thick and strong, overhanging roofs shade all the openings, shutters cover the windows. It is a habitation designed for a hot climate with frequent flooding. (p. 109)

11. "In the case of change due to disasters, Philips and Stukes (2003) argue that 'disasters bring about a renegotiation of place, throwing location and identity into question'" (p. 17). So, a disaster occurs and affects the place ties that define the self and community (Philips & Stukes, 2003). Amends to that disaster are proposed and debated within those disrupted frameworks of place attachment while, as Alkon (2004) notes,

"people's ideas about themselves and their daily lives (here, in relation to place) mediate specific political decisions" (p. 148).

12. Reflexive inclusion also involves the appreciation on the part of government officials of the local pubic sphere's history, ethnic-gender composition, and culture in relation to past and present power relationships that have motivated negative consequences that compromise the development of trust (Miller & Rivera, 2006a).

13. Schneider (1995) noted the following:

> Bureaucratic norms develop slowly and methodically inside public organizations; they set the parameters for acceptable governmental activity. Emergent norms originate instantaneously and spontaneously within a disaster-stricken population, providing the framework for individual and social behavior. . . . The problem is that these two sets of norms may evolve in very different directions. When this occurs, there is a noticeable difference or gap between governmental plans and the needs of the affected population. So the size of the gap captures the degree of harmony or discord that exists between bureaucratic and emergent norms in any given disaster situation. (pp. 55-56)

14. Dye and Zeigler (1996, as cited in Anderson, 2006) summarize the elite theory of public policy with the following seven points:

(1) "Society is divided into the few who have power and the many who do not";

(2) "The few who govern are not typical of the masses who are governed. Elites are drawn disproportionately from the upper socioeconomic strata of society";

(3) "The movement of non-elites to elite position must be slow and continuous to maintain stability and avoid revolution. Only non-elites who have accepted the basic elite consensus can be admitted to governing circles";

(4) "Elites share consensus on the basic values of the social system and the preservation of the system";

(5) "Public policy does not reflect demands of the masses but rather the prevailing values of the elite";

(6) "Elites may act out of narrow self-serving motives and risk undermining mass support, or they may initiate reforms, curb abuse, and undertake public-regarding programs to preserve the system and their place in it"; and

(7) "Active elites are subject to relatively little direct influence from apathetic."

Bibliography

ABC News. (2005, September 6). *Who's to blame for delayed response to Katrina?* Retrieved November 6, 2005, from ABCNews.go.com.

Aberbach, J. D., & Walker, J. L. (1970). Political trust and racial ideology. *The American Political Science Review, 64*, 1199-1219.

Abney, F. G., & Hill, L. B. (1966). Natural disasters as a political variable: The effect of a hurricane on an urban election. *The American Political Science Review, 60,* 974-981.

Aday, D., & Ito, S. (1989). Social structure and disaster: A prolegomenon. In G. Kreps (Ed.), *Social structure and disaster* (pp. 19-26). Newark, DE: University of Delaware Press.

Adeola, F. (1998). Environmental justice in the state of Louisiana: Hazardous wastes and environmental illness in the cancer corridor. *Race, Gender & Class, 6,* 83.

Alcubilla, M. (1891). *Diccionario de la administracion espanola* (5th ed.) Madrid, Spain: Administración.

Alive in Truth. (2005). Kevin L., interview September 20, 2005. Retrieved May 29, 2006, from http: //www.aliveintruth.org/stories/Kevin_L.htm.

Alkon, A. H. (2004). Place, stories, and consequences: Heritage narratives, the control of erosion on Lake County, California vineyards. *Organization and Environment, 17,* 145-169.

Altman, I., & Low, S. M. (1992). Place attachment. In I. Altman & S. M. Low (Eds.), *Place attachment.* New York: Plenum Press.

Anderson, J. E. (2006). *Public policymaking: An introduction* (6th ed.). New York: Houghton Mifflin Company.

Aptekar, L. (1990). A comparison of the bicoastal disasters of 1989. *Behavior Science Research, 24*(1-4), 73-104.

Arata, C., Picou, J. S., Johnson, G. D., & McNally, T. S. (2000). Coping with technological disaster: An application of the conservation of resources model to the *Exxon Valdez* oil spill. *Journal of Traumatic Stress, 13*(1), 23-39.

Associated Press. (2005). Toxic mold spreads through soggy south. Retrieved January 1, 2007, from http://www.msnbc.msn.com/id/9505817/.

Atkins, D., & Moy, E. M. (2005). Left behind. The legacy of Hurricane Katrina: Hurricane Katrina puts the health effects of poverty and race in plain view. *British Medical Journal, 331,* 916-918.

Audio Recordings of Hurricane Katrina Conference Calls. (2005a). LA State Emergency Operations Center. August 26-28.

Avery, J. M. (2006). The sources and consequences of political mistrust among African Americans. *American Politics Research, 34,* 653-682.

Backhaus, G. (2005). Introduction. In G. Backhaus & J. Murungi (Eds.), *Lived topographies and their mediational forces* (pp. xi-xxix). Lanham, MD: Lexington Books.

Baert, P. (1998). *Social theory in the twentieth century.* Cambridge, UK: Polity Press.

Baker, J. K. (2003). *Landscapes: Nature, culture and the production of space.* Pittsburg, PA: University of Pittsburgh.

Barber, B. (1983). *The logic and limits of trust.* New Brunswick, NJ: Rutgers University Press.

Barry, J. M. (1997). *Rising tide: The Great Mississippi Flood of 1927 and how it changed America.* New York: Simon & Schuster.

Barton, A. H. (1970). *Communities in disaster.* Garden City, NY: Anchor/Doubleday.

Basso, K. H. (1996). *Wisdom sits in places.* Albuquerque, New Mexico: University of New Mexico Press.
Bates, F., Fogleman, C. W., Parenton, B., Pittman, R., & Tracy, G. (1963). *The social and psychological consequences of a natural disaster.* Washington, DC: National Academy of Sciences.
Baumgartner, F. R., & Jones, B. D. (1993). *Agendas and instability in American politics.* Chigago: University of Chicago Press.
Bea, K. (2005a, September 2). *Disaster evacuation and displacement policy: Issues for Congress.* Cong. Res. Serv., Order No. RS 22235.
Bea, K. (2005b, September 2). *Louisiana emergency management and homeland security authorities summarized.* Cong. Res. Serv., Order No. RL 32678.
Bea, K., Runyon, L. C., & Warnock, K. M. (2004, March 17). *Emergency management and homeland security statutory authorities in the states, District of Columbia, and insular areas: A summary.* Cong. Res. Serv. Order No. RL 32287.
Beck, V. (1992). *Risk society: Towards a new modernity.* Thousand Oaks, CA: SAGE Publications.
Berger, P., & Luckmann, T. (1967). *The social construction of reality: A treaties on the sociology of knowledge.* New York: Anchor Books.
Berke, P. (1998). Reducing natural hazard risks through state growth management. *Journal of American Planning Association, 64,* 76-88.
Berke, P. R., & Campanella, T. J. (2006). Planning for post disaster resiliency. *The Annals of The American Academy of Political & Social Science, 604,* 192-207.
Berube, A., & Katz, B. (2005). *Katrina's window: Confronting concentrated poverty across America.* Washington, DC: The Brookings Institution.
Birch, E. (2006). Learning from past disasters. In E. Birch & S. M. Wachter (Eds.), *Rebuilding urban places after disaster: Lessons from Hurricane Katrina* (pp. 132-148). Philadelphia: University of Pennsylvania Press.
Bird, E. S. (2002). It makes sense to us. *Journal of Contemporary Ethnography, 31,* 519-547.
Birkland, T. A. (1996). Natural disasters as focusing events: Policy communities and political response. *International Journal of Mass Emergencies and Disasters, 14,* 221-243.
Birkland, T. A. (1997). *After disaster: Agenda setting, public policy, and focusing events.* Washington, DC: Georgetown University Press.
Blanco, K. B. (2005, August 27). *Letter from Kathleen Babineaux Blanco, Governor of LA, to George W. Bush, President of the United States.* Retrieved March 16, 2007, from http://www.gpoaccess.gov/katrinareport/mainreport.pdf.
Bohannan, P. (1995). *How culture works.* New York: The Free Press.
Bolding, G. (1969). The New Orleans seaway movement. *Louisiana History, 10*(1), 49-60.
Bolin, R. (1986). Disaster impact and recovery: A comparison of black and white victims. *International Journal of Mass Emergencies and Disasters,* 4(1), 35-50.
Bolin, R., & Bolton, P. A. (1986). *Race, religion, and ethnicity in disaster recovery.* Monograph No. 42. Boulder, CO: University of Colorado, Institute of Behavioral Science.
Bonnes, M., & Secchiaroli, G. (2003). *Introduction to environmental psychology. Lecture four: Place attachment and place identity.* Retrieved June 20, 2006, from http://sss-student-tees.ac.uk/psychology/modules/year2/environmental_psych/lect4.doc.
Bourdieu, P. (1980). Le capital social: Notes provisoires. *Actes de la Recherche en Sciences Sociales, 3,* 2-3.
Bourdieu, P. (1983). Forms of capital. In J. G. Richardson (Ed.), *Handbook of theory and research for the sociology of education* (pp. 241-258). New York: Greenwood Press.

Bourdieu, P. (1983/1986). The forms of capital. In J.G. Richardson (Ed.), *Handbook of theory and research for the sociology of education* (pp. 241-258). Westport, CT: Greenwood Press.

Bow, V., & Buys, L. (2003, November 21). *Sense of community and place attachment: The natural environment plays a vital role in developing a sense of community.* Presented to the Social Change in the 21st Century Conference. Queensland University of Technology.

Bowser, B. A. (2006, April 4). Restoring Louisiana wetlands. Retrieved June 15, 2006, from http://www.pbs.org/newshow/bb/science/jan-jun06/wetlands_4-3.html.

Brader, T. (2005). Striking a responsive chord: How political ads motivate and persuade voters by appealing to emotions. *American Journal of Political Science, 49,* 388-405.

Brader, T. (2006). *Campaigning for hearts and minds: How emotional appeals in political ads work.* Chicago, IL: University of Chicago Press.

Brasseaux, C. A. (1995). The image of Louisiana and the failure of voluntary French Emigration, 1683-1731. In J. K. Schafer, W. M. Billings, & G. R. Conrad (Eds.), *The Louisiana Purchase bicentennial series in Louisiana history* (Vol. 1: The French Experience in Louisiana). Lafayette, LA: University of Southwest Louisiana Studies.

Breunlin, R., & Regis, H. A. (2006). Putting the Ninth Ward on the map: Race, place and transformation in Desire, New Orleans. *American Anthropologist, 108,* 744-764.

Breunlin, R. (2004). *Space swapping: Urban renewal and public housing.* Retrieved March 3, 2004, from http://www.endsound.com/nopf/socialcomments/spaceswapping.htm.

Brinkley, D. (2006). *The great deluge: Hurricane Katrina, New Orleans and the Mississippi Gulf Coast.* New York: Morrow.

Bronson, W. (1996). *The earth shook, the sky burned.* San Francisco: Chronicle Books.

Brookings Institute, The. (2005). *New Orleans after the storm: Lessons from the past, a plan for the future.* Washington, DC: Brookings Institution.

Brown, B. B., & Perkins, D. D. (1992). Disruptions in place attachment. In I. Altman & S. M. Low (Eds.), *Place attachment* (pp. 279-304). New York: Plenum Press.

Brown, M. (2006). Disturbed rest. *The Times-Picayune.* New Orleans, LA, A:1, A10-A11.

Buchholz, R. (1993). *Principles of environmental management: The greening of business.* Englewood Cliffs, NJ: Prentice Hall.

Bullard, R. D. (1990a). *Dumping in Dixie: Race, class and environmental quality.* Boulder, CO. Westview Press.

Bullard, R. D. (1990b). Ecological inequalities and the new south: Black communities under siege. *Journal of Ethnic Studies, 17,* 101-115.

Bullard, R. D. (2000). *Dumping in Dixie: Race, class and environmental quality.* (3rd Ed.). Boulder, CO. Westview Press.

Burby, R. J. (2006). Hurricane Katrina and the paradoxes of government disaster policy: Bringing about wise governmental decisions for hazardous areas. *The Annals of The American Academy of Political & Social Science, 604,* 171-191.

Burby, R. J., Beatley, T., Berke, P. R., Deyle, R. E., French, S. P., Godschalk, D. R., et al. (1999). Unleashing the power of planning to create disaster-resistant communities. *Journal of the American Planning Association, 65,* 247-258.

Bureau of the Census. (2004). *American community survey.* Retrieved January, 12, 2007, from http://www.census.gov/acs/www/.

Burley, D., Jenkins, P., & Azcona, B. (2006). Loss, attachment, and place: Land loss and community coastal Louisiana. *Community and Ecology: Dynamics of Place, Sustainability and Politics Research in Urban Policy, 10,* 21-42.

Burt, R. S. (1992). *Structural holes: The social structure of competition.* Cambridge, MA: Harvard University Press.

Bush, R. K., Portnoy, J. M., Saxon, A., Terr, A. I., & Wood, R. A. (2006). The medical effects of mold exposure. *Journal of Allergy Clinical Immunology, 117,* 326-333

Buttimer, A. (1980). Social space and the planning of residential areas. In A. Buttimer & D. Seamon (Eds.), *The human experience of space and place* (pp. 21-54). New York: St. Martin's Press.

Butz, D., & Eyles, J. (1997). Reconceptualizing senses of place: Social relations, ideology and ecology. *Geografiska Annaler: Series B, Human Geography, 79,* 1-25.

Calvin, W. (1994). The emergence of intelligence. *Scientific America, 271,* 101-107.

Campanella, R. (2002). *Time and place in New Orleans: Past geographies in the present day.* Gretna, LA: Pelican Publishing Company.

Campanella, R. (2006a). *Geographies of New Orleans.* Lafayette, LA: University of Louisiana at Lafayette.

Campanella, R. (2006b). Reperceiving place. *Louisiana Cultural Vistas, 17*(2), 60-65.

Carmines, E. G., & Stimson, J. A. (1993). On the evolution of political issues. In W. H. Riker (Ed.), *Agenda formation* (pp. 151-168). Ann Arbor, MI: The University of Michigan Press.

Carr, L. (1932). Disaster and the sequence-pattern concept of social change. *The American Journal of Sociology, 38,* 207-218.

Carrns, A. (2005, November 25). *Long before flood, New Orleans system was prime for leaks.* Retrieved November 27, 2005, from http://www.online.wsj.com/article/SB1132882717410006253.html.

Casey, E. (1987). *Remembering: A phenomenological study.* Bloomington, IN: Indiana University Press.

Castree, N., & Braun, B. (2001). *Social nature: Theory, practice and politics.* Oxford, UK: Blackwell.

Centers for Disease Control and Prevention. (2006). Health concerns associated with mold in water-damaged homes after Hurricane Katrina and Rita—New Orleans area, Louisiana, October 2005. *MMWR: Morbidity Mortal Weekly Report, 55*(2), 41-44.

Chafe, W. H., Gavins, R., & Korstad, R. (2001). *Remembering Jim Crow: African Americans tell about life in the segregated South.* New York: Russell Sage.

Chapman, W. (1979). *Preserving the past.* London, UK: Dent.

Cigler, B. (2005, October 31). *FEMA, Homeland Security may need a divorce.* Retrieved November 7, 2005, from http://www.centredaily.com.

City of New Orleans. (2006). *Rebuilding New Orleans.* Retrieved January, 21, 2007, from http://www.bringneworleansback.org/Portals/BringNewOrleansBack/Resources/Mayors%20Rebuilding%20Plan%20Final.pdf.

City Planning Commission, City of New Orleans. (1999). *New century New Orleans, 1999 Land Use Plan, city of New Orleans.* New Orleans: Author.

Clark, J. G. (1967). New Orleans and the river: A study of attitudes and responses. Louisiana History 8.

Cline, I. (1900, November 16). *Special report on the Galveston Hurricane of September 8, 1900.* Monthly Weather Review. U.S. Weather Bureau. Washington, D.C., 372-374.

CNN Reports. (2005). *Katrina: State of emergency.* Kansas City, MO: Andrews McMeal Publishing.

CNN. (2004). Hurricane Charley: Two days later (transcript). *CNN Live Sunday.* Retrieved March 17, 2007, from http://transcripts.cnn.com/TRANSCRIPTS/0408/15/sun.02.html.

CNN. (2005, September 18). *A disturbing view from inside FEMA.* Retrieved November 10, 2005, from http://www.cnn.com.

CNN.com. (2005, September 2). *Military due to move in to New Orleans.* Retrieved May 30, 2007, from www.cnn.com/2005/weather/09/02/Katrina.impact/index.html.
Cochrane, T. (1987). Place, people, and folklore: An Isle Royale case study. *Western Folklore, 46,* 1-20.
Cohen, C., & Werker, E. (2004, April). *Toward an understanding of the root causes of forced migration: The political economy of 'natural' disasters.* Working paper for the Mellon-MIT Inter-University Program on Non-Governmental Organizations and Forced Migration. No. 25. Retrieved March 10, 2007, from http://web.mit.edu/cis/www/migration/pubs/rrwp/25_towards.pdf.
Colby, K. M. (1981). Modeling a paranoid mind. *Behavioral and Brain Sciences, 4,* 515-560.
Coleman, J. S. (1988). Social capital in the creation of human capital. *American Journal of Sociology, 94*(Suppl.), S95-S120.
Coleman, J. S. (1990). *Foundations of social theory.* Cambridge, MA: Harvard University Press.
Comfort, L. (2006). Cities at risk: Hurricane Katrina and the drowning of New Orleans. *Urban Affairs Review, 41,* 501-516.
Comptroller of the Currency. (1976, August 31). *Cost, schedule, and performance problems of the Lake Ponchartrain and vicinity, Louisiana, Hurricane Protection Project.* PSAD-76-161. U.S. General Accounting Office. Washington, D.C.: General Accounting Office.
Congleton, R. D. (2006). The story of Katrina: New Orleans and the political economy of catastrophe. *Public Choice, 127,* 5-30.
Conover, P. J., & Feldman, S. (1986). Emotional reactions to the economy: I'm mad as hell and I'm not going to take it anymore. *American Journal of Political Science, 30,* 50-78.
Cook, K. S., & Cooper, R. M. (2003). Experimental studies of cooperation, trust, and social exchange. In E. Ostrom & J. Walker (Eds.), *Trust and reciprocity: Interdisciplinary lessons from experimental research* (pp. 209-244). New York: Russell Sage Foundation.
Cooper, C. (2005, September 8). *Old-line families escape worst of flood and plot future.* Retrieved July 4, 2006, from http://www.commondreams.org/headlines05/0908-09.htm.
Corcoran, M., & Chaudry, A. (1997). The dynamics of childhood poverty. *The Future of Children: Children and Poverty, 7*(2), 40-54.
Cordasco, K., Eisenman, D., Glick, D., Golden, J., & Asch, S. (2007). They blew the levee: Distrust of authorities among Hurricane Katrina evacuees. *Journal of Health Care for the Poor and Underserved, 18,* 277-282.
Cornell University. (2004). *Disability status reports: Louisiana.* Ithaca, NY: Rehabilitation Research and Training Center on Disability Demographics and Statistics.
Cosgrove, D. (1989). Geography is everywhere: Culture and symbolism in human landscapes. In D. Gregory & R. Walford (Eds.), *Horizons in human geography* (pp. 118-135). New York: Palgrave Macmillan Publishers.
Cowen, T. (2006, April 19). *An economist visits New Orleans: Bienvenido, Nuevo Orleans.* Retrieved January 10, 2007, from http://www.mercatus.org/Publications/pubID.2272,cfilter.0/pub_detail.asp.
Cross, M. E. (trans.). (1938). *Father Louis Hennipen's description of Louisiana: Newly discovered to the southwest of New France by order of the King.* Minneapolis: University of Minnesota Press.
Crouse, N. (2001). *Lemoyne D'Iberville: Soldier of New France.* Baton Rouge, LA: Louisiana State University Press.

Cuthbertson, B. H., & Nigg, J. M. (1987). Technological disaster and the nontherapeutic community: A question of true victimization. *Environment and Behavior, 19,* 462-483.

Cutler, D., & Glaeser, E. (1997). Are ghettoes good or bad? *Quarterly Journal of Economics, 112,* 827-872.

Cutter, S. L., & Emrich, C. T. (2006). Moral hazard, social catastrophe: The changing face of vulnerability along the hurricane coasts. *The Annals of the American Academy of Political & Social Science, 604,* 102-112.

Cutter, S. L., Boruff, B. J., & Shirley, W. L. (2003). Social vulnerability to environmental hazards. *Social Science Quarterly, 84,* 242-261.

Cutter, S. L., Emrich, C. T., Mitchell, J. T., Boruff, B. J., Gall, M., Schmidtlein, M. C., et al. (2006). The long road home: Race, class and recovery from Hurricane Katrina. *Environment, 48*(2), 8-20.

Cutter, S. L., Mitchell, J. T., & Scott, M. S. (2000). Revealing the vulnerability of people and places: A case study of Georgetown County, South Carolina. *The Annals of the Association of American Geographers, 90,* 713-737.

Cutter, S. L. (1996). Vulnerability to environmental hazards. *Progress in Human Geography, 20,* 529-539.

Dahl, R. A. (1961). *Who governs?* New Haven, CT: Yale University Press.

Daly, H. E., & Cobb, J. B. (1994). *For the common good: Redirecting the economy toward community, the environment, and a sustainable future.* Boston, MA: Beacon Press.

Dao, J. (2006, January 27). Study says 80 percent of New Orleans blacks may not return. *The New York Times,* A:18.

Davis, D. W. (2000). Historical Perspective on Crevasses, Levees, and the Mississippi River. In C. E. Colten (Ed.), *Transforming New Orleans and its environs: Centuries of change* (pp. 84-106). Pittsburg, PA: University of Pittsburgh Press.

Dawson, M. C. (2006). After the deluge: Publics and publicity in Katrina's wake. *Du Bois Review, 3,* 239-249.

Day, C. (2002). *Spirit & place: Healing our environment.* Oxford, England: Architectural Press.

De Bow, J. D. B. (1854). *Statistical view of the United States—Compendium of the seventh census.* Washington, DC: A.O.P. Nicholson, Printer.

De Villiers du Terrage, M. (1920). A history of the foundation of New Orleans, 1717-1722. *The Louisiana Historical Quarterly, 3,* 157-251.

Dean, C. (2005, October 4). *Some experts say it's time to evacuate the coast (for good).* Retrieved October 10, 2005, from http://www.nytimes.com.

Dean, C., & Reukin, A. (2005, August 31). Hurricane Katrina: Floodwaters; geography complicates levee repair. *New York Times,* A10.

Dean, M., & Walch, J. (1989). *The power of geography: How territory shapes social life.* Boston, MA: Unwin Hyman.

Dear, M. (1997). Postmodern bloodlines. In G. Beuko & U. Strohmayer (Eds.), *Space and social theory: Interpreting modernity and postmodernity* (pp. 49-71). Oxford, UK: Blackwell Publishers.

Dekaser, R. (2005). Economic consequences of Hurricane Katrina. *American Banking Association Banking Journal, 97*(11), 72.

Denhardt, R. B., & Glaser, M. A. (1999). Communities at risk: A community perspective on urban social problems. *National Civic Review, 88,* 145-153.

Deutsch, M. (1960). The effect of motivational orientation upon trust and suspicion. *Human Relations, 13,* 123-139.

Dirks, R. (1980). Social responses to severe food shortages and famine. *Current Anthropology, 21,* 21-44.

Dixon, J., & Dogan, R. (2003). Corporate decision making: Contending perspectives and their governance implications. *Corporate Governance: International Journal of Business in Society, 3,* 39-57.

Dixon, J., Dogan, R., & Sanderson, A. (2005). Community and communitarianism: A philosophical investigation. *Community Development Journal, 40,* 4-16.

Dominguez, V. R. (1986). *White by definition: Social classification in Creole Louisiana.* New Brunswick, NJ: Rutgers, The State University.

Dreier, P. (2006). Katrina and power in America. *Urban Affairs Review, 41,* 528-549.

Dugan, B. (2007). Loss of identity in disaster: How do you say goodbye to home? *Perspectives in Psychiatric Care, 43,* 41-46.

Duke, J. T. (1976). *Conflict and power in social life.* Provo, UT: Brigham Young University Press.

Duncan, O. D. (1961). From social system to ecosystem. *Sociological Inquiry, 31,* 140-149.

Dye, T., & Zeiglrt, L. H. (1996). *The irony of democracy: An uncommon introduction to American politics.* Belmont, CA: Wadsworth Publications.

Dynes, R. (1970). *Organizational behavior in disaster.* Lexington, MA: Heath-Lexington Books.

Dyson, M. E. (2006). *Come hell or high water: Hurricane Katrina and the color of disaster.* New York: Basic Civitas Books.

Ebbert, T. C. (2005, December 14). *Hearing on Hurricane Katrina: Preparedness and response by the state of Louisiana before select committee.* 109th Congress. Written statement of the Director of Homeland Security for New Orleans.

Edelstein, M. R. (2003). *Contaminated communities: Coping with residential toxic exposure* (2nd ed.). Boulder, CO: Westview Press.

Eeckhout, B. (2001). The 'Disneyification' of Times Square: Back to the future? In K. F. Gotham (Ed.), *Critical perspectives on urban redevelopment* (pp. 379-428). New York: Elsevier Press.

Ehrlich, P. (1994). Ecological economies and the carrying capacity of Earth. In A. Jansson, M. Hammer, C. Folke, & R. Costanza (Eds.), *Investing in natural capital: The ecological economies approach to sustainability* (pp. 38-56). Washington, DC: Island Press.

Ekberg, M. (2007). The parameters of the risk society: A review and exploration. *Current Sociology, 55,* 343-366.

Ellsworth, W. I. (1990). Earthquake history, 1769-1989. In R. E. Wallace (Ed.), *The San Andreas Fault System (Geological Survey Professional Paper 1515)* (pp. 152-187). Washington, DC: U.S. Government Printing Office.

Erickson, B. H. (1995, February). *Networks, success, and class structure: A total view.* Presented at the Sunbelt Social Networks Conference. Charleston, SC.

Erickson, B. H. (1996). Culture, class and connections. *American Journal of Sociology, 102,* 217-251.

Erickson, K. (1994). *A new species of trouble: The human experience of modern disasters.* New York. Norton.

Ethridge, R. (2006). Bearing witness: Assumptions, realities, and the otherizing of Katrina. *American Anthropologist, 108*(4), 799-813.

European Commission Joint Research Center. (n.d.) *Mapping severe damage to S.E. Asia's land cover following the Tsunami.* Retrieved April 30, 2007, from http://ies.jrc.cec.eu.int/fileadmin/Documentation/Reports/Global_Vegetation_Monitoring/Tsunami_Landcover_change.pdf.

Evans, B. (2006). *1906 San Francisco Earthquake Housing is Valuable Piece of History.* Retrieved April 27, 2006, from http://realitytimes.com/rtapages/20060418_quakehistory.htm.

Experts from Brown Hearing. (2005, September 27). Retrieved March 17, 2007, from http://online.wsj.com/article/SB112785315597053609.html.
Eyles, J. (1989). The geography of everyday life. In D. Gregory & R. Walford (Eds.), *Horizons in human geography* (pp. 102-117). Totowa, NJ: Barnes & Noble.
Falk, W. W., Hunt, M. O., & Hunt, L. L. (2006). Hurricane Katrina and New Orleans' sense of place, return and reconstruction or 'gone with the wind'? *Du Bois Review, 3*, 115-128.
Federal Emergency Management Agency. (2005). *By the Numbers: First 100 Days: FEMA Recovery Update for Hurricane Katrina.* Retrieved December 8, 2005, from http://www.fema.gov/media/archives/index120705.shtm.
Feldman, R. M. (1990). Settlement-identity: Psychological bonds with home places in a mobile society. *Environment and Behavior, 22*, 183-229.
Fiorino, D. J. (1990). Can problems shape priorities? The case of risk-based environmental planning. *Public Administration Review, 50*, 82-90.
Flap, H. D. (1991). Social capital in the reproduction of inequality. *Comparative Sociology of Family, Health and Education, 20*, 6179-6202.
Flap, H. D. (1994, July). *No man is an island: The research program of social capital theory.* Presented at the World Congress of Sociology. Bielefeld, Germany.
Florence, R. (1997). *New Orleans cemeteries: Life in the cities of the dead.* New Orleans: Batture Press.
Foote, K. E. (1988). Object as memory: The material foundations of human semiosis. *Semiotica, 69*, 259-263.
Foote, K. E. (1997). *Shadowed ground.* Austin, Texas: University of Texas Press.
Form, W. H., C. Loomis, R. Clifford, H. Moore, S. Nosow, G. Stone, and C. Westie. (1956). The persistence and emergence of social and cultural systems in disasters. *American Sociological Review, 21*(2), 180-185.
Freudenburg, W. R. (1993). Risk and recreancy: Weber, the division of labor, and the rationality of risk perceptions. *Social Forces, 71*, 909-932.
Freudenburg, W. R., & Jones, T. (1991). Attitudes and stress in the presence of technological risk: A test of the Supreme Court hypothesis. *Social Forces, 69*, 1143-1168.
Fried, M. (1963). Grieving for a lost home. In L.J. Duhl (Ed.), *The urban condition* (pp. 151-171). New York: Basic Books.
Fried, M. (1966). Grieving for a lost home: Psychological costs of relocation. In J. Q. Wilson (Ed.), *Urban renewal the record and the controversy* (pp. 359-379). Cambridge, MA: MIT Press.
Fritz, C. E. (1961). *Disasters and community therapy.* Washington, DC: National Research Council-National Academy of Science.
Frymer, P., Strolovitch, D. Z., & Warren, D. T. (2005, September 28). *Katrina's political roots and division: Race, class and federalism in American politics.* Retrieved December 28, 2005, from http://understandingKatrina.SSRC.org/FrymerStrolovitchWarren/.
Frymer, P., Strolovitch, D. Z., & Warren, D. T. (2006). New Orleans is not the exception: Re-politicizing the study of racial inequality. *Du Bois Review, 3*, 37-57.
Galvin, R. (2003). The Earth shook, the sky burned: The newspaper in American life. *Humanities, 24*(6), 50-54.
Garnett, J. L. (1992). *Communicating for results in government: A strategic approach for public managers.* San Francisco: Jossey-Bass Publishers.
General Accountability Office. (2005, July). *Homeland Security: DHS' efforts to enhance first responders' all hazard capabilities continue to evolve.* Pub. No. GAO-05-652.
Gibson, J. J. (1979). *The ecological approach to visual perception.* Boston, MA: Houghton Mifflin.

Giddens, A. (1984). *The construction of society: Outline of theory of structuration.* Berkeley, CA: University of California Press.

Giddens, A. (1989). A reply to my critics. In D. Held & J. B. Thompson (Eds.), *Social theory of modern societies: Anthony Giddens and his critics* (pp. 249-305). Cambridge, MA: Cambridge University Press.

Gieryn, T. F. (2000). A space for place in sociology. Annual Review of Sociology, 26, 463-496.

Gill, D. A. (1994). Environmental disaster and fishery co-management in a natural resource community: Impacts of the *Exxon Valdez* oil spill. In C. L. Dyer & J. R. Mcgoodwin (Eds.), *Folk management in the world's fisheries: Implications for fisheries managers* (pp. 207-235). Boulder, CO: University of Colorado Press.

Gill, D. A., & Picou, S. J. (1998). Technological disaster and chronic community stress. *Society and Natural Resources, 11,* 795-815.

Giraud, M. (1966). *Histoire de La Louisiane Francaise* (Vol. 3). Paris: Presses Universitaires de France.

Giroux, H. A. (2006). *Stormy weather: Katrina and the politics of disposability.* Boulder, CO: Paradigm Publishers.

Gladwin, T. N., Kennelly, J. J., & Krause, T.-S. (1995). Shifting paradigms for sustainable development: Implications for management theory and research. *The Academy of Management Review, 20,* 874-907.

Glaeser, E. (2005). Should government rebuild New Orleans, or just give residents checks? *Economists' Voice, 2*(4), Article 4.

Godkin, M. A. (1980). Identity and place: Clinical applications based on notions of rootedness and uprootedness. In A. Buttimer & D. Seamon (Eds.), *The human experience of space and place* (pp. 73-85). New York: St. Martin's Press.

Gold, J. R., & Ward, S.V. (1994). *Place promotion: The use of publicity and marketing to sell towns and regions.* New York: John Wiley and Sons.

Gordon, N. S., Farberow, N. L., & Maida, C. A. (1996). *Children and disasters.* New York: Brunner/Mazel.

Gormley, W. T., Jr., & Balla, S. J. (2004). *Bureaucracy and democracy: Accountability and performance.* Washington, DC: CQ Press.

Gothman, K. F. (2002). Marketing Mardi Gras: Commodification, spectacle and the political economy of tourism in New Orleans. *Urban Studies, 39,* 1735-1756.

Gottdiener, M. (1997). *Theming of America: Dreams, visions, and commercial spaces.* Bolder, CO: Westview Press.

Greene, M. (1992). *Housing recovery and reconstruction: Lessons from recent urban earthquakes.* In Proceedings of the 3rd U.S./Japan Workshop on Urban Earthquakes, Publication No. 93-B. Oakland, CA: Earthquake Engineering Research Institute.

Greider, T., & Garkovich, L. (1994). Landscapes: The social construction of nature and the environment. *Rural Sociology, 59,* 1-24.

Groat, L. (1995). Introduction: Place, aesthetic evaluation and home. In L. Groat (Ed.), *Giving places meaning* (pp. 1-26). New York: Academic Press.

Grogan, P., & Proscio, T. (2000). *Comeback cities: A blue print for urban neighborhoods renewal.* Boulder, CO: Westview.

Gross, E. (1995). *Somebody got drowned, Lord: Florida and the Great Okeechobee Hurricane disaster of 1928.* Doctoral dissertation, Florida State University.

Guillette, E. (1991). *The impact of recurrent disaster on the aged of Botswana.* Presentation at the 50th Annual Meeting of Social Applied Anthropology, Charleston, SC.

Gunter, V. & Kroll-Smith, S. (2007). *Volatile places: A sociology of communities and environmental controversies.* Thousand Oaks, CA: Pine Forge Press.

Habermas, J. (1992[1996]). *Between facts and norms: Contributions to a discourse theory of law and democracy* (trans. W. Rehg). Cambridge, MA: MIT Press.

Haddow, G. D., & Bullock, J. A. (2003). *Introduction to emergency management*. Boston, MA: Butterworth-Heinemann.
Hamerton, P. G. (1885). *Landscape*. Boston, MA: Roberts Bros.
Hanifan, L. J. (1916). The rural school community center. *Annals of the American Academy of Political and Social Science, 67*, 130-138.
Hannigan, J., & Kueneman, R. (1978). Anticipating flood emergence: A case study of a Canadian disaster subculture. In L. Quarantelli (Ed.), *Disasters: Theory and research* (pp. 129-146). Beverly Hills, CA: Sage Publications.
Hardin, R. (2002). *Trust and trustworthiness* (vol. 4). New York: Russell Sage Foundation.
Hardt, M., & Negri, A. (2004). *Multitude: war and democracy in an age of empire*. New York: Penguin Press H.C.
Harper, C. (1998). *Exploring social change* (3rd ed.). Upper Saddle River, NJ: Prentice Hall.
Harper, C. (2003). *Environment and society: Human perspectives on environmental issues* (3rd ed.). Upper Saddle River, NJ: Prentice Hall.
Harvey, D. (1989). *The condition of postmodernity: An enquiry into the origins of cultural change*. New York: Blackwell.
Hayden, D. (1996). *The power of place: Urban landscapes as public history*. Cambridge, MA: MIT Press.
Heidorn, K. (2000). *Dr. Isaac M. Cline. Part 2 - Converging paths: A man and a storm*. Retrieved May 1, 2007, from http://www.islandnet.com/~see/weather/history/icline2.htm.
Hewitt, K. (1983). *Interpretations of calamity: From the viewpoint of human ecology*. Boston, MA: Allen and Unwin.
Hodges, L. (2005, December 6). Testimony before the bipartisan committee to investigate the preparation for and response to Hurricane Katrina. Retrieved December 20, 2005, from http://katrina.house.gov/hearings/12_06_05/hodges_120605.rtf.
Holcomb, B. (1993). Revisioning place: De- and re-constructing the image of the industrial city. In G. Kearns & C. Philo (Eds.), *Selling places: The city as cultural capital, past and present* (pp. 133-144). Oxford, England: Pergamon Press.
Holling, C. S. (1994). New science and new investments for a sustainable biosphere. In A. Jansson, M. Hammer, C. Folke, & R. Costanza (Eds.), *Investing in natural capital: The ecological economics approach to sustainability* (pp. 57-73). Washington, DC: Island Press.
Hollis, M. (1994). *The philosophy of social science: An introduction*. Cambridge, UK: Cambridge University Press.
Holzer, H. J., & Lerman, R. I. (2006). Employment issues and challenges in post-Katrina New Orleans. In M. A. Turner & S. R. Zedlewski (Eds.), *After Katrina: Rebuilding opportunity and equity into the new New Orleans* (pp. 9-16). Washington DC: The Urban Institute.
Hosmer, L. T. (1994). Strategic planning as if ethics mattered. *Strategic Management Journal, 15*, 17-34.
Hossain, M. A., Ahmed, M. S., & Ghannoum, M. S. (2004). Attributes of Stachybotrys chartarum and its association with human disease. *Journal of Allergy and Clinical Immunology, 113*, 200-208.
Howlett, M. (1998). Predictable and unpredictable policy windows: Institutional and exogenous correlates of Canadian federal agenda-setting. *Canadian Journal of Political Science, 31*, 495-524.
Huigen, P. P., & Meijering, L. (2005). Making places: A story of DeVenen. In G. J. Ashworth & Graham, B. (Eds.), *Senses of place: Senses of time* (pp. 19-30). Burlington, VT: Ashgate Publishing Co.

Hutchinson, F. (2005). *Hurricane Katrina and the culture war: Can natural disasters change world views?* Retrieved January 10, 2007, from www.renewamerica.us/analysis/050906hutchinson.htm.

Hutter, M. (2007). *Experiencing cities*. Upper Saddle River, NJ: Pearson Education, Inc.

Hutter, M., Miller, D. S., & Rivera, J. D. (2006). *Street murals and community identity: A comparison of neighborhood art in Los Angeles, CA and Philadelphia, PA*. Presented at the 77th Annual Meeting of the Pacific Sociological Association, April 22-26, 2006, Hollywood, CA.

Ingold, T. (1992). Culture and the perception of the environment. In E. Croll & D. Parkin (Eds.), *Bush base: Forest farm–culture, environment and development* (pp. 39-56). New York: Routledge.

Jackson, M.-R. (2006). Rebuilding the cultural vitality of New Orleans. In M. A. Turner & S. R. Zedlewski (Eds.), *After Katrina: Rebuilding opportunity and equity into the new New Orleans* (pp. 55-61). Washington DC: The Urban Institute.

Jacobs, J. (1961). *The death and life of great American cities*. New York: Random House.

Jameson, F. (1991). Postmodernism, or, the cultural logic of late capitalism. Durham, NC: Duke University Press.

Janis, I. L. (1954). Problems of theory in the analysis of stress behavior. *The Journal of Social Issues, 10*, 12-125.

Jarvis, B. B., & Miller, J. D. (2005). Mycotoxins as harmful indoor air contaminants. *Applied Microbiology and Biotechnology, 66*, 367-372.

Jones, C. O. (1984). *An introduction to the study of public policy* (3rd ed.). Fort Worth, TX: Harcourt Brace College Publishers.

Kaslow, A. J. (1981). *Oppression and adaptation: The social organization and expressive culture of an Afro-American community in New Orleans, Louisiana*. New York: Columbia University.

Kastner, L., & Schirm, A. (2005). *State food stamp participation rates in 2003*. Washington, DC: Mathematica Policy Research Inc.

Katovich, M. A., & Couch, C. J. (1992). The nature of social pasts and their use as foundations for situated action. *Symbolic Interaction, 15*, 25-47.

Katovich, M. A., & Hintz, R. A., Jr. (1997). Responding to a traumatic event: Restoring shared pasts within a small community. *Symbolic Interaction, 20*, 275-290.

Katrina Timeline. (2005). *Think progress*. Retrieved March 16, 2007, from http://www.thinkprogress.org/katrina-timeline.

Katrina: What Happened When. (2005). Retrieved March 16, 2007, from http://www.factcheck.org/printerfriendly348.html.

Kelman, A. (2003). *A river and its city: The nature of landscape in New Orleans*. Berkeley, CA: University of California Press.

Kelsen, H. (1961). *General theory of law and state*. Translated by Anders Wedberg. New York: Russell & Russell.

Kephart, W. M. (1948). Is the American Negro becoming lighter? An analysis of the sociological and biological trends. *American Sociological Review, 13*(4), 437-443.

Kerwin, C. M. (1999). *Rulemaking: How government agencies write law and make policy* (2nd ed.). Washington, DC: Congressional Quarterly Inc.

Kettl, D. F. (2005). *The worst is yet to come: Lessons from September 11 and Hurricane Katrina—Report*. Philadelphia: University of Pennsylvania.

Kingdon, J. (1995). *Agendas, alternatives, and public policies* (2nd ed.). New York: HarperCollins College Publishers.

Klein, N. (2005, May 2). *The rise of disaster capitalism*. Retrieved June 12, 2005, from http://www.thenation.com/doc/20050502/klein.

Klein, N. (2006, August 30). *Disaster capitalism: How to make money out of misery.* Retrieved January 27, 2006, from http://www.guardian.co.uk/comment/story/0,,1860673,00.html.

Knox, V. W., London, A. S., Scott, E. K., & Blank, S. (2003). *Welfare reform, work, and child care: The role of informal care in the lives of low-income women and children.* New York: MDRC.

Kolb, C. (2006). Crescent city, post-apocalypse. *Technology and Culture, 47,* 108-111.

Korten, D. C. (1990). *Getting to the 21st century: Voluntary action and the global agenda.* West Hartford, CT: Kumarian Press.

Kramer, R. M. (2004). Collective paranoia: Distrust between social groups. In R. Hardin (Ed.), *Distrust* (vol. 8) (pp. 136-166). New York: Russell Sage Foundation.

Kramer, R. M., & Brewer, M. B. (1986). Social group identity and the emergence of cooperation in resource conservation dilemmas. In H. Wilke, C. Rutte, & D. M. Messick (Eds.), *Psychology of decisions and conflict: Experimental social dilemmas* (pp. 205-234). Frankfurt, Germany: Lang.

Kroll-Smith, J. S. (1995). 1994 MSSA plenary address: Toxic contamination and the loss of civility. *Sociological Spectrum, 15,* 377-396.

Kroll-Smith, J. S., & Couch, S. R. (1991). What is a disaster? An ecological symbolic approach to resolving the definitional debate. *International Journal of Mass Emergencies and Disasters, 9,* 355-366.

Kroll-Smith, J. S., & Couch, S. R. (1993a). Symbols, ecology and contamination: Case studies in the ecological-symbolic approach to disaster. *Research in Social Problems and Public Policy, 5,* 47-73.

Kroll-Smith, J. S., & Couch, S. R. (1993b). Technological hazards: Social responses as traumatic stressors. In J.P. Wilson & B. Raphael (Ed.), *International handbook of traumatic stress syndromes* (pp. 79-91). New York: Plenum Press.

Kroll-Smith, J. S., Couch, S. R., Marshall, B. K. (1997). Sociology, extreme environments and social change. *Current Sociology, 45*(3), 1-18.

Kyvig, D. E., & Marty, M. A. (2000). *Nearby history: Exploring the past around you (2nd ed).* Lanham, MD: Rowman & Littlefield Publishers, Inc.

La. Rev. Stat. 38.83.

Lang, R. E. (2006). Measuring Katrina's impact on the Gulf Megapolitan area. In E. L. Birch & S. M. Wachter (Eds.), *Rebuilding urban places after disaster: Lessons from Hurricane Katrina* (pp. 89-102). Philadelphia: University of Pennsylvania.

Larson, E. (1999). *Isaac's storm: A man, a time, and the deadliest hurricane in history.* New York: Crown Publishers.

Law.com. (2007). *Subrogation.* Retrieved March 3, 2007, from http://dictionary.law.com/default2.asp?review=true#hill.

Lefebvre, H. (1991). *The production of space.* (Trans. Donald Nicholson-Smith). Oxford, England: Basil Blackwell.

Levy, S. R., Freitas, A. L., Mendoza-Denton, R., & Kugelmass, H. (2006). Hurricane Katrina's impact on African Americans' and European Americans' endorsement of the Protestant work ethic. *Analyses of Social Issues and Public Policy, 6,* 75-85.

Lewis, P. (2003). *New Orleans: The making of an urban landscape* (2nd ed.). Santa Fe, NM: Center for American Places.

Lhulier, L., & Miller, D. S. (2006). Public relations, democracy and the establishment of effective organizational transparency. *International Journal of the Humanities, 3,* 242-249.

Lieberman, R. C. (2006). The storm that didn't discriminate: Katrina and the politics of color blindness. *Du Bois Review, 3,* 7-22.

Lin, N. (1982). Social resources and instrumental action. In P. V. Marsden & N. Lin (Eds.), *Social structure and network analysis* (pp. 131-145). Beverly Hills, CA: SAGE Publications.

Lin, N. (1995). Les resources sociales: Une theorie du capital social. *Revue Française de Sociologie, 36*, 685-704.

Lin, N. (2001). *Social capital: A theory of social structure and action.* Cambridge, UK: Cambridge University Press.

Lipsitz, G. (1988). Mardi Gras Indians: Carnival and counter-narrative in Black New Orleans. *Cultural Critique,* 10, 99-121.

Lotke, E., & Borosage, R. (2006). *Hurricane Katrina: Natural disaster, human catastrophe.* Washington, DC: Campaign for America's Future.

Loughlin, K. (1995). *Locating corporate environmentalism: The Bhopal case.* Presented at Advanced Seminar Power/Knowledge Shifts in America's Fin-de-Siecle, Santa Fe, NM.

Loury, G. (1977). A dynamic theory of racial income differences. In P. A. Wallace & A. LeMund (Eds.), *Women, minorities, and employment discrimination* (pp. 153-188). Lexington, MA: Lexington Books.

Lowenthal, D. (1985). *The past is a foreign country.* New York: Cambridge University Press.

Luhmann, N. (1988). Familiarity, confidence, trust: Problems and alternatives. In D. Gambetta (Ed.), *Trust* (pp. 94-107). Oxford, UK: Blackwell.

Lynch, K. (1960). *The image of the city.* Cambridge, MA: MIT Press.

MacGillivary, A., & Walker, P. (2000). Local social capital: Making it work on the ground. In S. Baron, J. Field, & T. Schuller (Eds.), *Social capital: Critical perspectives* (pp. 197-211). Oxford, UK: Oxford University Press.

Maclean, D. E. (1990). Comparing values in environmental policies: Moral issues and moral arguments. In P. B. Hammond & R. Coppock (Eds.), *Valuing health risks, costs and benefits for environmental decision making* (pp. 83-106). Washington, DC: National Academy Press.

Magill, J. (2003). New Orleans through three Centuries. In A. Lemmon, J. Magill, & J. R. Wiese (Eds.), *Charting Louisiana: Five hundred years of maps.* New Orleans, LA: The Historic New Orleans Collection.

Magill, J. (2005-2006). On perilous ground. *Louisiana Cultural Vistas, 16*(4), 32-45.

Magill, J. (2006). San Francisco, New Orleans and disasters of the centuries. *Louisiana Cultural Vistas, 17*(2), 46-57.

Maines, D. R., Sugrue, N.M., & Katovich, M. A. (1983). The sociological import of G. H. Mead's Theory of the Past. *American Sociological Review, 48,* 161-173.

Malone, L. A. (1990). *Environmental regulation of land use.* Deerfield, IL: Clark Boardman Callaghan.

Manuel, J. (2006). In Katrina's wake. *Environmental Health Perspectives, 114,* 32-39.

Marcus, G. E., & Mackuen, M. (1993). Anxiety, enthusiasm, and the vote: The emotional underpinnings of learning and involvement during presidential campaigns. *American Political Science Review, 87,* 672-685.

Marcus, G. E., Neuman, W. R., & MacKuen, M. (2000). *Affective intelligence and political judgment.* Chicago: University of Chicago Press.

Mariscal, J. (2003, May 30). Lethal and compassionate: The militarization of U.S. culture. Retrieved June 4, 2007, from http://www.counterpunch.org/mariscal10505 2003.html.

Marsalis, W. (2005). Saving American's soul kitchen. *Louisiana Endowment for the Humanities, 16*(3), 18-19.

Marshall, B., McQuaid, J., & Schleifstein, M. (2006, January 29). For centuries, canals kept New Orleans dry. Most people never dreamed they would become mother nature's instrument of destruction. *Times-Picayune* (New Orleans).

Maximus. (2002). *Comprehensive needs assessment of low-income families in Louisiana.* Fairfax, VA: Maximus.

McCully, P. (1996). *Silenced rivers: The ecology and politics of large dams.* London, UK: Zed Books.

McNabb, D. & Madère, L. E., Jr. (1983). *A history of New Orleans.* Retrieved December 15, 2006, from http://www.madere.com/history.html.

McPhee, J. (1989). *The control of nature.* New York: Farrar, Straus, and Giroux.

McQuaid, J., & Schleifstein, M. (2006). *Path of destruction: The devastation of New Orleans and the coming age of superstorms.* New York: Little, Brown and Company.

McSpadden, L. A. (1991). *Case management versus bureaucratic needs: Earthquake response in California.* Presented at 50th Annual Meeting of Social Applied Anthropology, Charleston, SC.

Mead, G. H. (1929). The nature of the past. In J. Coss (Ed.), *Essays in Honor of John Dewey* (pp. 235-242). New York: Henry Holt.

Mead, G. H. (1934). *Mind, self, and society.* Chicago, IL: University of Chicago Press.

Mehta, M. (1995). Environmental risks: A Macrosociological Perspective. In M. D. Mehta & E. Ouellet (Eds.), *Environmental sociology: Theory and practice* (pp. 185-202). North York: Captus Press.

Meiklejohn, A. (1948). *Political freedom: The constitutional powers of people.* New York: Harper.

Merton, R. K. (1969). *Foreword.* In A. H. Burton, *Communities in disaster: A sociological analysis of collective stress situations* (pp. vii-xxxvii). Garden City, NY. Doubleday.

Michelman, F. I. (1998). Brennan and democracy. *California Law Review, 86,* 399-427.

Miller, D. S. (2006a). *The aesthetic value of landscapes and place orientation after a natural disaster.* A paper presentation at the Pacific Sociological Association, Universal City, Los Angeles, CA.

Miller, D. S. (2006b). Visualizing the corrosive community: Looting in the aftermath of Hurricane Katrina. *Space and Culture, 9,* 71-73.

Miller, D. S., Gill, D. A., & Picou, J. D. (2000). Assigning blame: The interpretation of social narratives and environmental disasters. *Southeastern Sociological Review, 1,* 13-31.

Miller, D. S., & Rivera, J. D. (2006a). Guiding principles: Rebuilding trust in government and public policy in the aftermath of Hurricane Katrina. *Journal of Public Management & Social Policy, 12,* 37-47.

Miller, D. S., & Rivera, J. D. (2006b). Hallowed ground, place and culture: The cemetery and the creation of place. *Space and Culture, 9,* 334-350.

Miller, D. S., & Rivera, J. D. (2007a). Landscapes of disaster and place orientation in the aftermath of Hurricane Katrina. In D. L. Brunsma, D. Overfelt, & S. Picou (Eds.), *The sociology of Katrina: Perspectives on a modern catastrophe* (pp 141-154). Lanham, MD: Rowman & Littlefield.

Miller, D. S., & Rivera, J. D. (2007b). Setting the stage: Roots of social inequality and the human tragedy of Hurricane Katrina. In R. S. Swan & K. A. Bates (Eds.), *Through the eye of Katrina: Social justice in the United States.* Durham, NC: Carolina Academic Press.

Miller, D. S., Rivera, J. D., & Yelin, J. C. (forthcoming). Civil liberties: The line dividing environmental protest and eco-terrorists. *Journal for the Study of Radicalism.*

Miller, M. (1992). Gifts: 20 great ideas for teaching sociology. Life chances exercise. *Teaching Sociology, 20,* 316-320.

Milligan, M. J. (1998). Interactional past and potential: The social construction of place attachment. *Symbolic Interaction, 21*(1), 1-33.

Mitchell, J. K. (1999). *Crucibles of hazard: Mega-cities and disasters in transition.* Tokyo, Japan: UNU Press.

Mittler, E. (1988). Agenda-setting in nonstructural hazard mitigation policy. In L. K. Comfort (Ed.), *Managing disaster: Strategies and policy perspectives* (pp. 86-107). Durham, NC: Duke University Press.

Moore, H. E. (1964). *And the winds blew.* Austin, TX: University of Texas.

Moore, W. E. (1974). *Social change* (2nd ed.). Englewood Cliffs, NJ: Prentice Hall.

MSNBC. (2005, September 9). *AP Reporter Mark Humphrey.* Washington, DC: The Washington Post Company.

Mulcahy, M. (2005). *Hurricanes, poverty, and vulnerability: An historical perspective.* Retrieved September 22, 2005, from http://understandingkatrina.ssrc.org/Mulcahy.

Mulrine, A. (2005). Lots of blame. Retrieved September 19, 2005, from www.usnews.com.

Muniz, B. (2006). *In the eye of the storm: How the government and private response to Hurricane Katrina failed Latinos.* Washington, DC: National Council of La Raza.

Myers, L. (2005, December 7). *Were the levees bombed in New Orleans?* Retrieved December 20, 2005, from http://www.msnbc.msn.com/id/10370145.

National Aeronautics and Space Administration. (2005). Hurricane Season 2005: Katrina. Retrieved December 30, 2006, from http://www.nasa.gov/vision/earth/lookingatearth/h2005_katrina.html.

National Aeronautics and Space Administration. (2005). Retrieved June 20, 2007, from http://visibleearth.nasa.gov/view_rec.php?id=20263.

National Guard. (2005, December 7). *Overview of significant events Hurricane Katrina.*

National Oceanic & Atmospheric Administration. (2004). *Galveston storm of 1900.* Retrieved May 1, 2007, from http://www.history.noaa.gov/stories_tales/cline2.html.

National Weather Service. (2005). Hurricane Katrina. Retrieved September 25, 2007, from http://www.nhc.noaa.gov/archive/2005/pub/al122005.public.020.shtml?

New Orleans Times-Picayune, The. (1965). September 22.

Nigg, J. M., & Mileti, D. (2002). Natural hazards and disasters. In R. Dunlap & W. Michelson (Eds.), *Handbook of environmental sociology* (pp. 272-294). Westport, CT: Greenwood Press.

Nolan, B. (2005, July 24). *In storm, N.O. wants no one left behind.* Retrieved February 9, 2006, from http://www.nola.com/printer/pirnter.ssf?/base/news-10/112218456019803 0.xml.

Norberg-Schulz, C. (1980). *Genius loci: Towards a phenomenology of architecture.* New York: Rizzoli.

O'Driscoll, B. R., Hopkinson, L. C., & Denning, D. W. (2005). Mold sensitization is common amongst patients with severe asthma requiring multiple hospital admissions. *BMC Pulm Med, 5*(1): 4.

Ogasawara, H. (1999). *Living with natural disasters: narratives of the Great Kanto and the Great Hanshin earthquakes.* Doctoral dissertation, Northwestern University.

Oglesby, C. (2005). *Living in camp cemetery. CNN.com* as cited in *CNN Reports. Katrina: State of Emergency.* Kansas City, MO: McMeel Publishing.

Olasky, M. (2007). *The politics of disaster: Katrina, big government, and a new strategy for future crises.* Nashville, TN: Thomas Nelson..

Oliver-Smith, A. (1986). Introduction. Disaster context and causation: An overview of changing perspectives in disaster research. In A. Oliver-Smith (Ed.), *Natural disas-*

ters and cultural responses (pp. 1-34). Williamsburg, VA: College of William and Mary.

Oliver-Smith, A. (1996). Anthropological research on hazards and disasters. *Annual Review of Anthropology, 25,* 303-328.

Oliver-Smith, A. (2005). Communities after catastrophe: Reconstructing the material, reconstituting the social. In S. E. Hyland (Ed.), *Community building in the twenty-first century* (pp. 45-70). Santa Fe, NM: School of American Research Press.

Oliver-Smith, A., & Hoffman, S. M. (2002). Introduction: Why anthropologists should study disasters. In S. M. Hoffman & A. Oliver-Smith (Eds.), *Catastrophe and culture: An anthropology of disaster* (pp. 3-22). Santa Fe, NM: School of American Research Press/James Currey Ltd.

Olsen, M. E. (1978). *The process of social organization* (2nd ed.). New York: Holt Rinheart and Wilson.

Olson, E. D. (2005, October 6). The environmental effects of Hurricane Katrina. Hearing statement before Committee on Environment and Public Works of the United States Senate.

Orr, D. W. (1994). *Earth in mind: On education, environment, and the human prospect.* Washington, DC: Island Press.

Pabis, G. S. (2000). Subduing nature through engineering: Caleb G. Forshey and the levees-only policy, 1851-1881. In C. E. Colten (Ed.), *Transforming New Orleans and its environs: Centuries of change* (pp. 64-83). Pittsburg, PA: University of Pittsburgh Press.

Panel on Civic Trust and Citizen Responsibility. (1999). *A government to trust and respect: rebuilding citizen-government relations for the 21st century.* Washington, DC: National Academy of Public Administration.

Parker, K. (2005, September 14). Three heroes outwitted bureaucracy. *New Hampshire Union Leader.*

Parsons, T. (1961). An outline of the social system. In T. Parsons, E. A. Shills, K. Naegele, & J. R. Pitts (Eds.), Theories of society (vol. 1) (pp. 30-79). New York: Free Press.

Parsons, T. (1966). *Societies.* Englewood Cliffs, NJ: Prentice Hall.

Pastor, M., Bullard, R. D., Boyce, J. K., Fothergill, A., Morello-Frosch, R., & Wright, B. (2006). *In the wake of the storm: Environment, disaster, and race after Katrina.* New York: The Russell Sage Foundation.

Pastor, M., Bullard, R. D., Boyce, J. K., Fothergill, A., Morello-Frosch, R., & Wright, B. (2006). *In the wake of the storm: Environment, disaster, and race after Katrina.* New York: Russell Sage Foundation.

Pauchant, T. C., & Fortier, I. (1990). Anthropocentric ethics in organizations, strategic management and the environment: A topology. In P. Shrivastava and R. B. Lamb (Eds.), *Advances in strategic management* (vol. 6, pp. 99-114). Greenwich, CT: JAI Press.

Paxton, P. (1999). Is social capital declining in the United States? A multiple indicator assessment. *The American Journal of Sociology, 105,* 88-127.

Pelling, M. (2003). *The vulnerability of cities: Natural disasters and social resilience.* Sterling, VA: Earthscan Publications Ltd.

Pettit, K., & Kingsley, G. T. (2003). *Concentrated poverty: A change in course.* Washington, DC: Urban Institute.

Petto, C. (2007). *When France was king of cartography: The patronage and production of maps in early modern France.* Lanham, MD: Lexington Books.

Philips, B., & Stukes, P. A. (2003, August). *Freedom hill is not for sale: Resistance to mitigation buyouts in Princeville, North Carolina.* Paper presented at the Annual Meeting of the American Sociological Association, Atlanta, GA.

Phillips, B. D. (1993). Cultural diversity in disasters: Sheltering, housing, and long-term recovery. *International Journal of Mass Emergencies and Disasters, 11*(1), 99-110.

Phillips, S. (2005). *What went wrong in hurricane crises?* Interview transcript.

Picou, J. S., Marshall, B. K. (2007). Katrina as a paradigm shift: Reflections on disaster research in the twenty-first century. In D. L. Brunsma, D. Overfelt, & S. Picou (Eds.), *The sociology of Katrina: Perspectives on a modern catastrophe* (pp. 1-20). Lanham, MD: Rowman & Littlefield.

Picou, J. S., Marshall, B. K., & Gill, D. A. (2004). Disaster, litigation, and the corrosive community. *Social Forces, 82,* 1493-1522.

Platt, R. (2000, June 7-9). Extreme natural events: Some issues for public policy. Presented at Extreme Events Workshop, Boulder, CO.

Platt, R. H., & Rubin, C. B. (1999). Stemming the losses: The quest for hazard mitigation. In R. H. Platt (Ed.), *Disasters and democracy: The politics of extreme natural events* (pp. 69-107). Washington, DC: Island Press.

Poland, C. (2006). *When the big one strikes again: Estimated losses due to a repeat of the 1906 San Francisco Earthquake.* Retrieved April 15, 2007, from http://adsabs.harvard.edu/abs/2006AGUFMGC43A..01K.

Popkin, S. J., Turner, M. A., & Burt, M. (2006). Rebuilding affordable housing in New Orleans: The challenge of creating inclusive communities. In M. A. Turner & S. R. Zedlewski (Eds.), *After Katrina: Rebuilding opportunity and equity into the new New Orleans* (pp. 17-26). Washington, DC: The Urban Institute.

Portes, A. (1998). Social capital: Its origins and applications in modern sociology. *Annual Review of Sociology, 22,* 1-24.

Post, R. (1998a). Democracy, popular sovereignty, and judicial review. *California Law Review, 86,* 429-443.

Post, R. (1998b). Introduction: After Bakke. In R. Post & M. Rogin (Eds.), *Race and representation: Affirmative action* (pp. 13-28). New York: Zone Books.

Post, R. (2006). Democracy and equality. *American Academy of Political & Social Sciences, 603,* 24-36.

Press Secretary, Office of the White House. (2005, September 15). *Fact sheet: President Bush addresses the nation on recovery from Katrina.* Retrieved March 17, 2007, from http://www.whitehouse.gov/news/releases/2005/09/20050915-7.html.

Proshansky, H. M., Fabian, A. K., & Kaminoff, R. (1983). Place-identity: Physical world socialization of the self. *Journal of Environmental Psychology, 3,* 57-83.

Putnam, R. D. (1993). *Making democracy work: Civic traditions in modern Italy.* New York: Basic Books.

Putnam, R. D. (1995). Bowling alone: America's declining social capital. *Journal of Democracy, 6,* 65-78.

Putnam, R. D. (2000). *Bowling alone: The collapse and revival of American community.* New York: Touchstone.

Quinones, S. (2006, April 4). *Migrants find a gold rush in New Orleans.* Retrieved January 27, 2007, from http://www.latimes.com/news/nationworld/nation/la-na-labor4apr04,0,5618062.story?page=1&coll=la-home-headlines.

Radelat, A. (2005, October). *An avalanche of aid: Hispanic workers flood the Gulf Coast.* Retrieved November 24, 2006, from www.nclr.org/files/36812_file_WP_Katrina_FNLfnl.pdf.

Ratanasermpong, S., & Polngam, S. (2006, September 18-22). *Application of geoinformatics for natural disasters management in Thailand.* Seventeenth United Nations Regional Cartographic Conference for Asia and the Pacific. Bangkok, Thailand.

Read, P. (1996). *Returning to nothing: The meaning of lost places.* New York: Cambridge University Press.

Reichl, A. (1997). Historic preservation and progrowth politics in U.S. cities. *Urban Affairs Review, 32*, 513-535.
Reichl, A. (1999). *Reconstructing Times Square: Politics and culture in urban development*. Lawrence, KS: University Press of Kansas.
Relph, E. (1976). *Place and placelessness*. London, UK: Pion.
Rice, A. (2005). Do you know what it means to lose New Orleans? *Louisiana Cultural Vistas, 16*(3), 12-15.
Richard, C. E. (2003). *Louisiana: An illustrated history*. Baton Rouge, LA: The Foundation for Excellence in Louisiana Public Broadcasting.
Riley, R. B. (1992). Attachment to the ordinary landscape. In I. Altman & S. M. Low (Eds.), *Place attachment* (pp. 13-36). New York: Plenum Press.
Ripley, A., Tumulty, K., Thompson, M., & Carney, J. (2005, September 11). 4 places where the system broke down. Retrieved March 22, 2007, from http://www.time.com/time/magazine/article/0,9171,1103560,00.html.
Ritchie, L. A. (2004). *Voices of Cordova: Social capital in the wake of the Exxon Valdez oil spill (Alaska)*. Doctoral dissertation, Mississippi State University.
Ritchie, L. A., & Gill, D. A. (2007). Social capital theory as an integrating theoretical framework in technological disaster research. *Sociological Spectrum, 27*, 103-129.
Ritea, S. (2006, January 3). *La. Won't Run N.O. Schools by Itself*. Retrieved February 15, 2006, from http://www.nola.com/news/t-p/metro/index.ssf?/base/news-12/1136270 23252410.xml.
Rivera, J. D., & Miller, D. S. (2006). A brief history of the evolution of United States' natural disaster policy. *Journal of Public Management and Social Policy, 12*, 5-14.
Rivera, J. D., & Miller, D. S. (2007). Continually neglected: Situating natural disasters in the African American experience. *Journal of Black Studies, 37*, 502-522.
Robbins, C. A., Swenson, L. J., Nealley, M. L., Gots, R. E., & Kelman, B. J. (2000). Health effects of mycotoxins in indoor air: A critical review. *Applied Occupational and Environmental Hygiene, 15*, 773-784.
Roberts, C. E. (2003). *Louisiana: An illustrated history*. Baton Rouge, LA: The Foundation for Excellence in Louisiana Public Broadcasting.
Robichaux, Albert J., Jr. (1973). *Louisiana census and militia lists 1770-1789 (vol. 1): German coast, New Orleans, below New Orleans and Lafourche*. Harvey, LA: Dumag Printing.
Robinson, Walter. (2005). *Hurricane Katrina and the Arts*. Retrieved January, 20, 2007, from http://www.artnet.com/magazineus/news/robinson/robinson9-15-05.asp.
Roig-Franzia, M. (2005, September 21). Ghost town: Down at the end of the world, houses walk and the dead rise up. *Washington Post*, C1.
Roig-Franzia, M., & Connolly, C. (2005, November 29). Night and day in New Orleans. Retrieved December 20, 2006, from http://www.washingtonpost.com/wp-dyn/content/article/2005/11/28/AR2005112801681_pf.html.
Rolston, H., III. (1988). *Environmental ethics: Duties to and values in the natural world*. Philadelphia: Temple University Press.
Rosenstone, S. J., & Hansen, J. M. (1993). *Mobilization, participation, and democracy in America*. New York: Macmillan.
Rossi, P. H., Wright, J. D., Weber-Burdin, E., & Pereira, J. (1983). *Victims of the environment: Loss from natural hazards in the United States, 1970-1980*. New York: Plenum Press.
Rousseau, J.-J. (1800). *Contract Social ou Principes du Droit Politique*. Paris, France: Garnier Freres.
Ruscio, K. P. (1996). Trust, democracy, and public management: A theoretical argument. *Journal of Public Administration Research and Theory, 6*, 461-477.

Sack, R. D. (1992). *Place, modernity, and the consumer's world: A relational framework for geographic analysis*. Baltimore, MD: John Hopkins University Press.
Saenz, R. (2005). *Beyond New Orleans: The social and economic isolation of urban African Americans. Population Reference Bureau*. Retrieved October 26, 2005, from http://www.prb.org/template.cfm?section=PRB&template=content/contentgroups/05_Articles/The_Social_And_Economic_Americans.htm.
Saito, Y. (1998). Appreciating nature on its own terms. *Environmental Ethics, 20*, 135-149.
Schneekloth, L. H., & Shibley, R. G. (1995). *Placemaking: The art and practice of building communities*. New York: John Wiley & Sons, Inc.
Schneider, S. K. (1995). *Flirting with disaster: Public management in crisis situations*. New York: M.E. Sharpe.
Scott, J. C. (1985). *Weapons of the weak*. New Haven, CT: Yale University Press.
Searles, H. F. (1960). *The nonhuman environment*. New York: International University Press.
Selznick, P. (2002). *The communitarian persuasion*. Washington, DC: Woodrow Wilson Center Press.
Sen, A. (1982). Approaches for the choice of discount rates for social benefit-cost analyses. In R. C. Lind (Ed.), *Discounting for time and risk in energy policy* (pp. 325–353). Washington, DC: Resources for the Future.
Shallat, T. (2006). Holding Louisiana. *Technology and Culture, 47*, 102-107.
Short, J .R. (1999). Urban imagineers: Boosterism and the representations of Cities. In A. E. G. Jonas & D. Wilson (Eds.), *The urban growth machine: Critical perspectives two decades later* (pp. 37-54). Albany, NY: State University of New York Press.
Shughart, W. F., II. (2006). Katrinanomics: The politics and economics of disaster relief. *Public Choice, 127*, 31-53.
Silver, C. S., & DeFries, R. S. (1990). *One Earth/one future: Our changing global environment*. Washington, DC: National Academy Press.
Silverstein, M. (1992). *Disasters: Your right to survive*. New York: Brassey's.
Simon, J. G. (1981). *The ultimate resource*. Princeton, NJ: Princeton University Press.
Slivka, J. (2005, September 12). Another flood that stunned America. *US News and World Report*, p. 26.
Smith, J. P., & Kington, R. S. (1997). Race, socioeconomic status, and health late in life. In L. G. Martin & B. Soldo (Eds.), *Racial and ethnic differences in the health of older Americans* (pp. 53-94). Washington, DC: National Academy Press.
Sobel, R. S., & Leeson, P. T. (2006). Government's response to Hurricane Katrina: A public choice analysis. *Public Choice, 127*, 55-73.
Solomon, G. M., Hjelmroos-Koski, M., Rotkin-Ellman, M., & Hammond, S. K. (2006). Airborne mold and endotoxin concentrations in New Orleans, Louisiana, after flooding, October through November 2005. *Environmental Health Perspectives, 114*, 1381-1386.
Somers, D. (1974). Black and White in New Orleans: A study in urban race relations, 1865-1900. *The Journal of Southern History, 40*(1), 19-42.
Sorkin, M. (1992). *Variations on a theme park: The new American city and the end of public space*. New York: Hill and Wang.
Sortable List of Dissimilarity Scores. (2005). Albany, NY: Lewis Mumford Center, SUNY-Albany. Retrieved January 2, 2007, from http://mumford.albany.edu/census/WholePop/WPsort/sort_d1.html.
Sprin, A. W. (1998). *The language of landscape*. New Haven, CT: Yale University Press.
Starik, M. (1995). Should trees have managerial standing? Toward stakeholder status for non-human nature. *Journal of Business Ethnics, 14*, 207-218.

Stark, P. C., Burge, H. A., Ryan, L. M., Milton, D. K., & Gold, D. R. (2003). Fungal levels in the home and lower respiratory tract illnesses in the first year of life. *American Journal of Respiratory Critical Care Medicine, 168,* 232-237.

Stead, W. E., & Stead, J. G. (1992). *Management for a small planet: Strategic decision making and the environment.* Thousand Oaks, CA: SAGE Publications.

Stein, A., & Preuss, G. B. (2006). Oral history, folklore, and Katrina. In C. Hartman & G. D. Squires (Eds.), *There is no such thing as a natural disaster* (pp. 37-58). New York: Routledge.

Steinberg, T. (2006). *Acts of God: The unnatural history of natural disaster in America* (2nd ed.). New York: Oxford University Press.

Steinbrugge, K. (1982). *Earthquake, volcanoes, and tsunamis: An anatomy of hazards.* New York: Skandia America Group.

Strom, E. (1999). Let's put on a show! Performing acts and urban revitalization in Newark, New Jersey. *Journal of Urban Affairs, 21.* 423-435.

Studs, T. (1968). *Division street: America.* Allen Lane, London: Penguin Press.

Sylves, R. T. (2006). President Bush and Hurricane Katrina: A presidential leadership study. *The Annals of The American Academy of Political & Social Science, 604,* 26-56.

Thomas, E., Gegax, T. T., Campo-Flores, A., Murr, A., Meadows, S., Darman, J., et al. (2005, September 19). *Katrina: How Bush blew it.* Retrieved March 16, 2007, from http://www.msnbc.msn.com/id/9287434/site/newsweek/.

Throop, G. M., Starik, M., & Rands, G. P. (1993). Sustainable strategy in a greening world: Integrating the natural environment into strategic management. In P. Shrivastava, A. Huff, & J. E. Dutton (Eds.), *Advances in strategic management* (vol. 9, pp. 63-92). Greenwich, CT: JAI Press.

Tibbetts, J. (2006, January). Louisiana's wetlands: A lesson in nature appreciation. *Environmental Health Perspectives, 114,* A40-A43.

Tidwell, M. (2006). *The ravaging tide: Strange weather, future Katrinas, and the coming death of America's coastal cities.* New York: Free Press, Simon & Schuster, Inc.

Tierney, K. J. (2005) *Recent developments in U.S. Homeland Security policies and their impacts for the management of extreme events.* Boulder, CO: University of Colorado at Boulder.

Times-Picayune. (2005, September 7). *Timeline of the levee breach.* Retrieved January 6, 2006, from http://www.indybay.org/newsitems/2005/09/07/17575571.php.

Times-Picayune Staff. (2006). Katrina: The ruin and recovery of New Orleans. *Times-Picayune.* New Orleans, LA.

Torry, W. I. (1979). Anthropological studies in hazardous environments: Past trends and new horizons. *Current Anthropology, 20,* 517-541.

Trainer, P., & Hutton, J. (1972). An approach to the differential distribution of deaths and disaster. Paper presented at the meeting of the Midwest Council on Social Research in Aging, Kansas City, KS.

Transcript: New Orleans' Mayor Ray. C. Nagin's Interview. (2005, September 2). Interview with WWL-AM. Retrieved July 20, 007, from http://www.jacksonfreepress.com/comments.php%id=7051_0_44_0_c.

Treasury Department. (1889). *Tables showing arrivals if alien passengers and immigrants in the United States from 1820 to 1888.* Washington, DC. Government Printing Office.

Tregle, J. G., Jr. (1992). Creoles and Americans. In A. R. Hirsch & J. Logsdon (Eds.), *Creole New Orleans: Race and Americanization* (pp. 131-185). Baton Rouge, LA: Louisiana State University Press.

Tuan, Y-F. (1977). *Space and place: The perspective of experience.* Minneapolis, MN: University of Minnesota Press.

Tuan, Y-F. (1991). 'It happened not too far from here...': A survey of legend theory and characterization. *Western Folklore, 49*, 371-390.

Twigg, D. K. (2004). *The winds of change? Exploring political effects of Hurricane Andrew*. Dissertation of Political Science at Florida International University.

United Nations Development Programme. (1994). *Human Development Report 1994*. New York: Author.

U.S. Bureau of Labor Statistics. (2005, November 4). Employment situation summary: October 2005. Retrieved January 27, 2007, from http://www.bls.gov/news.release/empsit.nr0.htm.

U.S. Department of Homeland Security. (2004, June 30). *Final draft: National response plan*. Retrieved March 16, 2007, from http://www.ema.ohio.gov/PDFs/NRP_Final_Draft.pdf.

U.S. Department of the Interior. (1997). *Environmental atlas of the Lake Ponchartrain basin*. Retrieved February 17, 2007, from http://pubs.usgs.gov/of/2002/of02-206/intro/toc.html.

U.S. General Accountability Office. (2004, May 13). *Emergency preparedness: Federal funds for first responders*. GAO-04-788T. Washington, DC: GAO.

U.S. General Accountability Office. (2005, April 12). *Homeland security: Management of first responder grant programs and efforts to improve accountability continue to evolve*. Statement of William O. Jenkins, Jr., Director of Homeland Security and Justice. GAO-05-530T. Washington, DC: GAO.

U.S. Geological Survey. (2006). The great 1906 San Francisco earthquake. Retrieved September 24, 2007, from http://earthquake.usgs.gov/regional/nca/1906/18april/index.php.

U.S. House of Representatives. (2006). *A failure of initiative*. Retrieved March 16, 2007, from http://www.gpoaccess.gov/katrinareport/mainreport.pdf.

U.S. Weather Service. (2006). *Memorial web page for the 1928 Okeechobee Hurricane*. Retrieved January 3, 2006, from http://www.srh.noaa.gov/mfl/newpage/Okeechobee.html.

Urgen Weather Message. (2005, August 28). *National Weather Service, New Orleans, Louisiana*, Retrieved June 5, 2006, from www.srh.noaa.gov/data/warn_archive/LIX/NPW/0828_155101.txt.

Uslaner, E. M. (2001). Volunteering and social capital: How trust and religion shape civic participation in the United States. In P. Dekker & E. M. Uslaner (Eds.), *Social capital and participation in everyday life* (pp. 104-117). London, UK: Routledge.

Uslaner, E. M., & Brown, M. (2005). Inequality, trust, and civic engagement. *American Politics Research, 3*, 868-894.

Vago, S. (1999). *Social change* (4th ed.). Upper Saddle River, NJ: Prentice Hall.

Vale, L., & Campenalla, T. (2005). *The resilient city: How modern cities recover from disaster*. New York: Oxford University Press.

Van Orden, D. (2002). *Race and natural disaster victims: A critical analysis of the Great Storm of 1928*. Philadelphia: Temple University.

Walker, H. J., & Derro, R. A. (1990). In F. B. Kniffen, *Cultural diffusion and landscapes: Selections by Fred B. Kniffen (Geoscience and Man, vol. 27)* (pp. 153-187). Baton Rouge, LA: Geoscience Publications, Louisiana State University.

Waugh, W. L., Jr. (2005). Katrina, Rita, and all-hazards emergency management. *Journal of Emergency Management, 3*, 1-2.

Waugh, W. L., Jr. (2006). The political costs of failure in the Katrina and Rita disasters. *The Annals of The American Academy of Political & Social Science, 604*, 10-25.

Waugh, W. L., Jr., & Sylves, R. T. (2002). Organizing the war on terrorism. *Public Administration Review, 62*, 145-153.

Webster, R. A. (2005, November 7). Storm clouds on Treme's future. Retrieved November 12, 2006, from http://www.neworleanscitybusiness.com/viewStory.cfm?recID= 13847.
Wegener, D. T., & Petty, R. E. (1998). The naïve scientist revisited: Naïve theories and social judgment. *Social Cognition, 16,* 1-7.
Weiss, E. B. (1989). *In fairness to future generations: International law, common patrimony, and intergenerational equity.* Dobbs Ferry, NY: The United Nations University and Transnational Publishers.
Welch, M. R., Rivera, R. E. N., Conway, B. P., Yonkoski, J., Lupton, P. M., & Giancola, R. (2005). Determinants and consequences of social trust. *Sociological Inquiry, 75,* 453-473.
Wenkart, A. (1961). Regaining identity through relatedness. *American Journal of Psychoanalysis, 21,* 227-233.
Werger, D., & Weller, J. M. (1973). *Disaster subcultures: The cultural residues of community disasters.* Columbus, OH: Columbus Disaster Research Center, Ohio State University.
Williams, D. R., & Patterson, M. E. (1996). Environmental meaning and ecosystem management: Perspectives from environmental psychology and human geography. *Society and Natural Resources, 9,* 507-521.
Williams, D. R., Patterson, M. E., & Roggenbuck, J. W. (1992). Beyond the commodity metaphor: Examining emotional and symbolic attachment to place. *Leisure Science, 14,* 29-46.
Williams, L. (2006). Quote page 26. In The Times-Picayune Katrina: The Ruin and Recovery of New Orleans. Times-Picayune Staff. New Orleans, LA.
Williams, P. D., & Vaske, J. J. (2002). *The measurement of place attachment: Validity and generalizability of a psychometric approach* (pp.1-28). Paper written for the U.S. Department of Agriculture, Forrest Services, Rocky Mountain Research Station.
Wilson, B. M. (1980). Social space and symbolic interaction. In A. Buttimer & D. Seamon (Eds.), *The human experience of space and place* (pp. 153-147). New York: St. Martin's Press.
Wilson, S., Jr. (1968). *The Vieux Carre New Orleans: Its plan, its growth, its architecture. Vieux Carre Historic Diotric Demonstration Study* (vol. 7). New Orleans, LA: The Bureau of Government Research.
Wood, B. D., & Doan, A. (2003). The politics of problem definition: Applying and testing threshold models. *American Journal of Political Science, 47,* 640-653.
Wood, G. S., Jr., & Judikis, J. C. (2002). *Conversations on community theory.* West Lafayette, IN: Purdue University Press.
World Bank. (1990). *World development report 1990: Poverty.* New York: Oxford University Press.
World Bank. (2000). *World development report 2000.* Oxford, UK: Oxford University.
Wright, B., Bryant, P., & Bullard, R. D. (1994). Coping with poisons in cancer alley. In R. D. Bullard (Ed.), *Unequal protection: Environmental justice and communities of color* (pp. 110-129). San Francisco: Sierra Club Books.
Wuthnow, R. (1998). *Loose connections: Joining together in America's fragmented communities.* Cambridge, MA: Harvard University Press.
Wuthnow, R. (2002). The United States: Bridging the privileged and the marginalized? In R. D. Putnam (Ed.), *Democracies in flux: The evolution of social capital in contemporary society* (pp. 59-102). Oxford, UK: Oxford University Press.
Yamagishi, T. (2001). Trust as a form of social intelligence. In K. S. Cook (Ed.), *The Russell Sage Foundation Series on trust* (1st ed., pp. 121-147). New York: Russell Sage Foundation.

Ydstie, J. (2006, August 27). *Katrina victims still struggle to find way home: All things considered.* Retrieved December 12, 2006, from http://www.npr.org/templates/story/story.phb?storyID=5720114.

Yilmax, Y., Hoo, S., & Nagowski, M. (forthcoming). *Measuring the fiscal capacity of states: A representative revenue system/representative expenditure system approach.* Washington, DC: The Urban Institute.

Young, A. A., Jr. (2006). Unearthing ignorance: Hurricane Katrina and the re-envisioning of the urban Black poor. *Du Bois Review, 3,* 206-213.

Zandi, M., Cochrane, S., Ksiazkiewicz, F., & Sweet, R. (2006). Restarting the economy. In E. L. Birch & S. M. Wachter (Eds.), *Rebuilding urban places after disaster: Lessons from Hurricane Katrina* (pp. 103-116). Philadelphia: University of Pennsylvania.

Zedlewski, S. R. (2006a). Pre-Katrina New Orleans: The backdrop. In M. A. Turner & S. R. Zedlewski (Eds.), *After Katrina: Rebuilding opportunity and equity into the new New Orleans* (pp. 1-8). Washington, DC: The Urban Institute.

Zedlewski, S. R. (2006b). Building a better safety net for the new New Orleans. In M. A. Turner & S. R. Zedlewski (Eds.), *After Katrina: Rebuilding opportunity and equity into the new New Orleans* (pp. 63-72). Washington, DC: The Urban Institute.

Zerubavel, E. (1996). Social memories: Steps to a sociology of the past. *Qualitative Sociology, 19,* 283-299.

Zhang, L. (2002). Spatiality and urban citizenship in late socialist China. *Public Culture, 14,* 311-334.

Zukin, S. (1996). Space and symbols in an age of decline. In A. King (Ed.), *Representing the city: Ethnicity, capital, and culture in the 21st century metropolis* (pp. 43-59). New York: New York University Press.

Index

Adaptive upgrading, 96

Betrayal: equivocal, 121-122; premeditated, 121; structural, 121-122
Bienville, 25-27, 29-30, 43nn2-3. *See also* Le Moyne, Jean-Bapiste
Blanco, Kathleen, 3, 79-81, 83, 125n20
Bonnet Carre Spillay, 69

Colonization: See DeSoto; French rule, 24-27, 47; LaSalle; Le Moyne, Jean-Bapiste; Le Moyne, Pierre; Spanish possession, 27, 47; United States possession, 28, 47-48

Davis-Bacon Act of 1931, 57-58
Department of Homeland Security, 76-78, 82, 89n14, 89n18, 122
DeSoto, Hernando, 24
DHS. See Department of Homeland Security
Disaster capitalism, 57, 64
Disaster landscape, 3-5, 15, 20, 22, 103
Disaster Mitigation Act, 72
Disaster subculture, 142-143
Disasters, 4
   Physical, 2; Social, 2
Disneyfication, 63-64, 66n21

Environmental Protection Agency, 39, 89n13
EPA. *See* Environmental Protection Agency
Epistemological predispositions, 116, 118
Equivocal betrayal, 121-122
Escherichia coli, 38. *See also* toxins

Federal Emergency Management Agency, 3, 74-77, 87, 89n13, 89n19, 90n21, 119, 122-122
Federal Insurance Administration, 72
FEMA. *See* Federal Emergency Management Agency
FIA. *See* Federal Insurance Administration
Flood Control Act, 68

Galveston Hurricane, 20-21
Government Performance and Results Act, 75
Governor of Louisiana. *See* Kathleen Blanco

Iberville. 25-26, 43n3. *See also* Le Moyne, Pierre
Interactional past, 6, 15-17, 21, 126-127, 136, 143n1
Interactional potential, 6, 15-17, 21, 46, 49, 65, 126-128, 136-137, 143n1
Lake Pontchartrain Hurricane Protection Project, 70
LaSalle, 24, 25
Le Moyne, Pierre, 25
Le Moyne, Jean-Bapiste, 25, 27
Lefebvre, Henri, 12, 13
Lifeworlds, 51
Louisiana Department of Environmental Quality, 39
Lower 9th Ward, 2, 34-35, 37, 40, 50, 55, 93

Mardi Gras tribes, 56
Mayor of New Orleans. *See* Ray C. Nagin
Mississippi River Commission, 30, 68

Nagin, Ray C., 2, 34, 59-60, 79-83, 85, 93-94, 122
National Flood Insurance Reform Acts, 72
Negative effects theory, 85
New Orleans Department of Health, 39

Ontological predispositions, 116-118

Place attachment, 1-2, 5, 10, 14-17, 41, 49-51, 61, 64, 67, 92, 95, 98, 124, 126, 134-138, 143, 144n11
Place identity, 14
Place memory, 16,
Placemaking, 12, 14-15
Premeditated betrayal, 121

Reflexive inclusion, 134-135

San Francisco earthquake, 17-19
Social recreancy, 118
Southeast Asian Tsunami, 19-20, 33
Structural betrayal, 121-122
The Swampland Acts, 68

Toxins: analysis of, 39; *Escherichia coli*, 38; fecal coliform bacteria, 38; mold, 40-41; toxic gumbo, 39; "toxic soup," 38;

Urban landscape, 9, 16, 27, 56, 12, 131

About the Authors

DeMond Shondell Miller is an associate professor of sociology, research scientist, and the director of the Liberal Arts and Sciences Institute for Research and Community Service at Rowan University in Glassboro, New Jersey. He has worked as principal investigator to facilitate research projects involving: environmental issues and community satisfaction. His primary areas of specialization are environmental sociology, disaster studies, the study of the social construction of place, community development, and social impact assessment. Dr. Miller has presented and published several professional papers; recent examples of such work can be found in: *The Researcher, The Qualitative Report, The Journal of Emotional Abuse, Space and Culture: An International Journal of Social Spaces, International Journal of the Humanities, Journal of Black Studies, The Journal for the Study of Radicalism, The Journal of Public Management and Social Policy, Sociological Spectrum,* and *The International Journal of Culture, Tourism and Hospitality Research.* Currently, he is heading a study focusing on the recovery process as survivors return to the post-Katrina Gulf Coast Region.

Jason David Rivera is a research associate at Rowan University in the Liberal Arts and Sciences Institute for Research and Community Service. He has recently worked on research dealing with public policy in reference to disaster mitigation and relief, social justice in the face of disasters, the reconfiguration of landscapes and their affect on local and global politics, and minority dynamics in civic engagement. Examples of Jason's research can be found in *The Journal of Public Management and Social Policy, Space and Culture: An International Journal of Social Spaces, Journal of Black Studies, The Journal for the Study of Radicalism,* and *International Journal of the Humanities.* Currently, he is studying the treatment of different American minority groups during natural disasters by the American government.